Praise for Kenneth C. Davis

Don't Know Much About® Mythology

"Ken Davis is the high school teacher we all wish we'd had—smart, funny, and irreverent. *Don't Know Much About® Mythology* is a crystallized reminder of what's enduring about the past, and why it continues to matter today. It's a perfect companion to Harry Potter, *The Da Vinci Code*, and the Bible—and the best excuse I know to get deserted on a Greek island. Between Odysseus and Icarus, it might even give you a few clues for how to find your way home—and how not to."
—Bruce Feiler, author of *Walking the Bible*

"Because Davis ranges widely and with such sparkling wit through a broad sweep of myths, his survey provides a superb starting point for entering the world of mythology."
—*Publishers Weekly* (starred review)

"An engaging handbook on gods, goddesses, and the civilizations they have inspired. . . . [Davis's] goal as an author is to infect readers with his own intellectual eagerness, and he succeeds admirably with this idiosyncratic tour of world mythology. . . . Even professors will have to concede that Davis has done his research—his annotated bibliography is excellent—and that he's a laudably conscientious scholar. An accessible and informed guide to an always-fascinating subject, and an ideal reference for the general reader."
—*Kirkus Reviews*

"A massive overview of every myth under the sun. Davis shatters commonly held myths about myths, differentiating them from allegories and legends, and explores the history of such tales in societies and religions around the globe, from Mesopotamia's Gilgamesh to Genesis' Noah. You can read here, too, about Native Americans' use of peyote, a tempestuous Nordic god of thunder, and a debate over the meaning of evil."
—*Daily News*

"With his trademark wit and fiercely entertaining style, Kenneth Davis draws us into mythological worlds, preserving ancient mysteries and enchantments even as he clarifies, orders, and makes sure we have the stories straight. *Don't Know Much About® Mythology* frames questions that arouse curiosity and produces answers that lead to astonishment. Whether you want a crash course on North American native myths or a refresher course on Gilgamesh, this book will provide a great read and remain a permanent reference manual."

—DR. MARIA TATAR, Department of Folklore and Mythology, Harvard University, and author of *The Annotated Brothers Grimm*

"In each of his *Don't Know Much About®* books, Kenneth C. Davis has brought the forgotten child to the front row, reminding those of us who hated school that one size doesn't fit all in education—that the desire to learn is far better served by the pursuit of individual passion than by classroom conformity. In *Don't Know Much About® Mythology*, Davis uses the intense passion that stirred in his own soul as a fifth-grade boy reading the *Odyssey*, to take us to a place of magic, imagination, and transcendence. Davis not only presents an entertaining exploration of humanity's most sacred stories across many civilizations, he brings us face to face with our most distant ancestors who were driven by innate curiosity to explain life's mysteries. Davis's book is a masterpiece. I couldn't stop turning the pages."

—ALBERT CLAYTON GAULDEN, founder and director of the Sedona Intensive and author of *Signs and Wonders*

"Kenneth C. Davis may have done more to educate our young people, and the general public, on the topics of history, geography, and science than all of the certified teachers in the country."

—*Orlando Sentinel*

Don't Know Much About® History—
THE BOOK THAT HAS SOLD MORE THAN 1.4 MILLION COPIES

"Don't Know Much About® History . . . puts the zest back in history."

—*Washington Post Book World*

"Quirky, sardonic, accurate, rudimentary, and often amusing. . . . A breezy question-and-answer approach that is far removed from the massive textbooks all of us once lugged around."

—*Atlanta Journal-Constitution*

"Unquestionably a handy reference book. It's just the thing when your pesky school-age kids try to show you up at homework time."

—*Dallas Morning News*

"Fun, engrossing, and significant. . . . History in Davis's hands is loud, coarse, painful, funny, irreverent—and memorable."

—*San Francisco Chronicle*

"Davis writes with humor, he can turn a fine phrase. . . . If history were usually taught this way, we wouldn't have to worry about the closing of the American mind."

—*Booklist*

"If you've always wondered exactly what Boss Tweed bossed and what Tammany Hall was, Davis is your man."

—*Washington Post Book World*

Don't Know Much About® Geography

"Davis pulls the reader west and east across the page with humor and a fresh viewpoint."

—*USA Today*

"Makes geography buffs of us all."

—*San Francisco Chronicle*

"Playful . . . fun. . . . His books will go a long way in helping people to answer that simple, important question: 'Where am I?'"

—*Chicago Tribune*

Don't Know Much About® the Universe

"Beware: You may easily become addicted to Davis's breezy style and his treasury of odd and fascinating facts."

—*Washington Post*

"Covers a huge topic in a lighthearted, conversational question-and-answer format."

—*Atlanta Journal-Constitution*

"A handy reference book, with a slew of good information and a host of age-old questions, intelligently arranged and amusingly addressed."

—*St. Louis Post-Dispatch*

Don't Know Much About® the Civil War

"Highly informative and entertaining. . . . Propels the reader light-years beyond dull textbooks and *Gone With the Wind*."

—*San Francisco Chronicle*

"Lively and relevant."

—*USA Today*

"Distinct, clear, and balanced. . . . Davis has a gift for deftly rendering the essentials."

—*New York Times Book Review*

Don't Know Much About® the Bible

"Do you still postpone reading the Bible cover to cover? Don't Know Much About® the Bible offers a rousing companion volume to get you going. Kenneth Davis will take you on a grand tour—with commentary."

—*Christian Science Monitor*

"A great starting point for Bible beginners."

—*San Francisco Chronicle*

"Reading him is like returning to the classroom of the best teacher you ever had!"

—*People*

"One of the most lucid, important, and worthwhile books of this or any year. Here is a perfect book for the dawn of the third millennium. Historian and Renaissance man Kenneth Davis is the perfect writer to tackle this lofty and compelling topic, which he does with zest and gusto, style and spirit, heart, humor, and integrity."

—Dan Millman, author of *Way of the Peaceful Warrior* and *Everyday Enlightenment*

"*Don't Know Much About® the Bible* is like a journey through biblical times with a modern tour guide who is fluent in history, psychology, religion, sociology, and anthropology! It's fascinating how Davis gives this ancient text such a rich development. I think his book would be wonderful to read to children."

—Carol Adrienne, author of *The Purpose of Your Life*

© Nina Subin

About the Author

KENNETH C. DAVIS, the *New York Times* bestselling author of *Don't Know Much About® History*, was recently dubbed the "King of Knowing" by Amazon.com. He often appears on national television and radio, and has served as a commentator on NPR's *All Things Considered*. In addition to his adult titles, he writes the Don't Know Much About® children's series, published by HarperCollins. He and his wife live in New York City and Vermont. They have two grown children, Jenny and Colin. Kenneth C. Davis is also the author of *America's Hidden History: Untold Tales of the First Pilgrims, Fighting Women, and Forgotten Founders Who Shaped a Nation*.

DON'T KNOW MUCH ABOUT®

anything else

Also by Kenneth C. Davis

America's Hidden History

THE DON'T KNOW MUCH ABOUT® SERIES

BOOKS FOR ADULTS

Don't Know Much About Anything

Don't Know Much About History

Don't Know Much About Geography

Don't Know Much About the Civil War

Don't Know Much About the Bible

Don't Know Much About the Universe

Don't Know Much About Mythology

BOOKS FOR CHILDREN

Don't Know Much About the 50 States

Don't Know Much About the Solar System

Don't Know Much About the Presidents

Don't Know Much About the Pilgrims

Don't Know Much About Mummies

Don't Know Much About Planet Earth

Don't Know Much About Space

Don't Know Much About American History

Don't Know Much About World Myths

Don't Know Much About Abraham Lincoln

Don't Know Much About Rosa Parks

Don't Know Much About Thomas Jefferson

EVEN MORE THINGS YOU NEED TO KNOW BUT NEVER LEARNED ABOUT PEOPLE, PLACES, EVENTS, AND MORE!

DON'T KNOW MUCH ABOUT® *anything else*

KENNETH C. DAVIS

HARPER

NEW YORK • LONDON • TORONTO • SYDNEY

HARPER

HarperCollins books may be purchased for educational, business, or sales promotional use. For information please write: Special Markets Department, HarperCollins Publishers, 10 East 53rd Street, New York, NY 10022.

FIRST EDITION

Library of Congress Cataloging-in-Publication Data is available upon request.

ISBN 978-0-06-156232-7

08 09 10 11 12 OV/RRD 10 9 8 7 6 5 4 3 2 1

This book is dedicated to
every child who ever asked "Why?"
And to every parent, teacher, librarian, friend, and relative
who might take the time to answer.

CONTENTS

INTRODUCTION

WHAT DO YOU WANT TO BE WHEN YOU GROW UP?

Adults have been putting kids on the spot with that one for a very long time. In fact, I still ask myself that question all the time and I am in my fifties! But do you remember what you wanted to be when you were young?

I happen to know because recently I came across a box of my own grade school papers stored away in the attic. Way back in elementary school, I wrote that perfunctory, "What I Want to Be" essay. Back then—and even into my college years—the idea that I could grow up and earn a living as a writer was as remote to me as being an astronaut or a neurosurgeon was. It simply wasn't on my radar. No, in my essay, I declared that I would be a teacher helping children to learn. But more surprisingly, even then, I declared that I would make my classroom more fun. My own teachers must have loved reading that!

So the notion that making education more interesting than it was for most of us may have been in my DNA. It was certainly on my mind back then in the William H. Holmes Elementary School in Mount Vernon, New York. That is where I usually found myself watching the clock on the wall or the construction vehicles across the street.

And it also seems apparent that I never thought that *fun* is a four-letter word. No, I've always held that *fun* and *education* can mingle, date, get serious. Even commit!

Confirmation of my long-held and admittedly idiosyncratic approach to learning came a few years ago, when I added children's books to my Don't Know Much About® series. At that time, I began making visits to schools. Wherever I travel, I take along a portable "Don't Know Much About Quiz Show," complete with buzzers and flashing lights. For an hour or so, I turn a classroom, cafeteria, or an auditorium into the set of *Who Wants to Be a Millionaire?* or *Jeopardy.*

The reaction among most students is the same, whether it is in the South Bronx, rural South Dakota, or suburban Connecticut. Kids are the same wherever I go—endlessly curious and they like to have fun. Everybody—and I mean everybody—is itching to get their fingers on one of those little buzzers. The excitement and enthusiasm of children who want to answer questions, show off what they know, and learn something new is an uncontainable force.

Yes, I ask some incredibly dorky questions, such as "What state is round at the ends and high in the middle? ("O-HI-O." Get it?) But that's part of the fun.

And wrong answers don't count. That's what makes some of the responses I get from the children so amusing. My favorite was the answer I got to the question, "What famous piece of clothing was invented for gold miners in California by Levi Strauss?" One little boy could barely contain himself and blurted out "Tights!"

I can't get the image of grizzled gold miners in tights out of my head.

After we finish playing the game, I let the students ask me questions about just about anything—from history to writing books to living in New York City. I have come to routinely expect two questions right up front, both of which reveal something about American values: "Are you rich?" and "Do you know anybody famous?"

But the truth is that almost everywhere I go, the excitement level and curiosity is much greater in elementary school than it is in the upper grades. Somewhere curiosity gets killed. It happens between sixth grade and high school. Clearly, a large number of the bright, interested—and infectiously curious—children leaping at the chance to learn are abducted by aliens and replaced by a race of bored, disaffected teenagers who hate school. Maybe it's just hormones. Or maybe it's the way we teach. But it doesn't have to be that way.

And that's why I used a column in *USA Weekend* magazine to create a series of quizzes on everything from major historical events, significant people, food, everyday inventions that have changed our lives, and extraordinary places. Many of them were sparked by my own curiosity. And in 2007, *Don't Know Much About Anything* collected more than 125 of those quizzes and became an immediate *New York Times* bestseller.

So like the proverbial bad penny, I'm back with a new collection of quizzes. I still have the same slightly twisted agenda: Ask some interesting questions, have some fun, and learn something together in the process. Once more, these quizzes are about Famous People (such as the woman who formed a bureau after the Civil War to search for missing men at Andersonville prison, or the First Lady who complained of "leading a dull life"); Exceptional Places (Which country has the most Catholics?); and Historic Happenings (Whose boot is immortalized in a statue at the Saratoga battlefield in upstate New York?); along with an array of questions about the worlds of sports, entertainment, and civics.

As I noted in the beginning of *Don't Know Much About Anything,* these quizzes cover a lot of intriguing territory. But they are not trivia. I believe that most of the information forms a compelling compendium of useful knowledge—even if it only helps you answer those troublesome *Jeopardy* questions when you play along at home.

You don't need a buzzer, but get ready to play!

FAMOUS PEOPLE

DON'T KNOW MUCH ABOUT *Clara Barton*

ON CHRISTMAS DAY, 1821, an "Angel" was born. Youngest of five children, Clara Barton was born in Oxford, Massachusetts. During the Civil War, she carried supplies to soldiers and nursed the wounded, for which she was called the "Angel of the Battlefield." Her legacy didn't end there as she founded the American Red Cross. Barton's thirty-eight-room home in Glen Echo, Maryland—where she died in 1912—was also the headquarters of the American Red Cross for several years and is now a National Historic site. What else do you know about this trailblazing "Angel" of mercy?

TRUE OR FALSE?

1. **Barton was a trained nurse.**
2. **When the war started, she worked in the U.S. Patent Office.**
3. **Barton was appointed head of nurses when the war started.**
4. **When the Civil War ended, she led the effort to locate missing soldiers.**
5. **The Red Cross was her idea.**
6. **The Red Cross originally assisted only soldiers.**

ANSWERS

1. False. Educated at home, she became a schoolteacher and her only medical experience was nursing an invalid brother, including applying leeches, a medical practice of the day.

2. True. She was the first female clerk in what is now the Patent and Trademark Office.

3. False. When the war broke out, she began a volunteer effort to carry medical supplies to the battlefield. But the government initially refused to give her help. In 1864, she was appointed superintendent of nurses for the Army of the James, a division of the Union forces.

4. True. Barton formed a bureau to search for missing men and the office marked more than 12,000 graves in the infamous Andersonville prison in Georgia.

5. False. In 1869, she went to Europe and learned of the International Committee of the Red Cross, based in Geneva. She took part in Red Cross activities during the Franco-Prussian War and when she returned to America campaigned for the establishment of the American branch of the Red Cross. She became its first president when it was founded in 1881. She also urged the United States government to ratify the Geneva Convention, which was done in 1882.

6. True. Barton was responsible for the clause in the Red Cross constitution that provides for civilian relief in peacetime calamities. It is known as the "American Amendment."

DON'T KNOW MUCH ABOUT
Salvador Dalí

"Every morning when I awake, the greatest of joys is mine: that of being Salvador Dalí." That outsized ego and showman's flair helped make Salvador Dalí one of the world's most recognizable artists. Born on May 11, 1904, in Figueres in Spain's Catalonian region, Dalí was an eccentric master whose abstract paintings, filled with bizarre images, are still extremely popular. Dalí died on January 23, 1989, in his hometown, where he is buried in the museum he built for himself. What do you know about this famed painter, whose wild mustache and love for the unusual were as famous as his art?

True or False?

1. **Dalí is best known for the style called "cubism."**
2. **Besides his paintings and lithographs, Dalí's most famous work is the notable short film *Un Chien Andalou.***
3. **A committed Communist, Dalí fought in the Spanish Civil War against General Franco.**
4. **Dalí collaborated with both Alfred Hitchcock and Walt Disney.**

Answers

1. False. Dalí's most famous work, including the painting *The Persistence of Memory,* is surrealist, which uses fantastic images to convey what Dalí called "hand-painted dream photographs." Cubism is usually associated with fellow Spaniard Picasso, whom Dalí greatly admired.

2. True. The sixteen-minute film made with Luis Buñuel is a surrealist classic with such shocking images as a woman's eyeball being sliced.

3. False. Briefly a Communist, Dalí left Spain during its Civil War and was later criticized when he returned to Spain and praised General Franco's repressive Fascist regime.

4. True. He worked on a dream sequence in Hitchcock's film *Spellbound* and collaborated with Disney on *Destino,* begun in 1945 and finally released in 2003.

DON'T KNOW MUCH ABOUT
Hans Christian Andersen

THE SON OF A poor shoemaker, born in Denmark on April 2, 1805, Hans Christian Andersen (1805–1875) became one of the world's most famous authors. After his father's death, the young boy set out to become a singer and actor, but had little luck until he turned to writing children's stories in 1835. His famed tales, often of luckless but goodhearted people—and animals—who find their fortune, have mesmerized readers ever since. What do you know about the man behind "The Ugly Duckling" and "The Little Mermaid"?

1. **In which story does a child say, "But the Emperor has nothing on"?**
2. **What really became of the Little Mermaid? Where is her famous statue located?**
3. **What is the test of "The Real Princess"?**
4. **What real person inspired Andersen to write "The Nightingale"?**
5. **Who played Andersen in the 1952 film based on his life and stories?**
6. **True or False? All of Andersen's 156 fairy tales are based on old folk tales.**

ANSWERS

1. "The Emperor's New Clothes."
2. In the original, she is turned into sea foam after her prince marries another girl. The statue, which has been vandalized on several occasions, is in Copenhagen's harbor.
3. She could feel three peas through twenty mattresses and twenty featherbeds.
4. The famed singer Jenny Lind, known as the "Swedish Nightingale," who was Andersen's great unrequited love.
5. Danny Kaye.
6. False. All but twelve are original stories.

DON'T KNOW MUCH ABOUT
President Eisenhower

ON JANUARY 20, 1953, World War II hero Dwight D. Eisenhower was inaugurated for the first of two terms as America's thirty-fourth President. He'd come a long way from his simple roots in Denison, Texas, where he was born on October 14, 1890, and Abilene, Kansas, where he grew up. West Point and an army career followed, bringing "Ike" to his pivotal role as the Allied commander of the D-Day invasion. His popularity after the war led to his election in 1952, and in 1953 he told America, "In the final choice, a soldier's pack is not so heavy a burden as a prisoner's chains." Eisenhower suffered a series of heart attacks and died of heart failure on March 28, 1969. Do you like Ike?

1. **Ike once served under a famous general he would later command. Who was he?**
2. **Before being elected U.S. President, what other presidency did Eisenhower hold?**
3. **In 1951, Eisenhower donned a uniform again. What did he command?**
4. **When he left the White House, Ike issued a famous but surprising warning. What danger did he foresee?**

ANSWERS

1. In 1933, he became an aide to General Douglas MacArthur, the Army chief of staff. In 1935, MacArthur became military adviser to the Commonwealth of the Philippines and took Eisenhower with him. After World War II, Ike was named Army chief of staff, making him MacArthur's superior.

2. In 1948, Eisenhower retired from active military service to become president of Columbia University in New York City.

3. He became chief of the North Atlantic Treaty Organization (NATO) in April 1951.

4. In a farewell speech, he said, "We must guard against the acquisition of unwarranted influence . . . by the military-industrial complex." Issued three days before Eisenhower left the White House in January 1961, the speech warned that "we must never let the weight of this combination endanger our liberties or democratic processes. We should take nothing for granted. Only an alert and knowledgeable citizenry can compel the proper meshing of the huge industrial and military machinery of defense with our peaceful methods and goals, so that security and liberty may prosper together."

DON'T KNOW MUCH ABOUT *Mother Cabrini*

ON JULY 7, 1946, Frances Xavier Cabrini (1850–1917) was declared a Catholic saint, the first U.S. citizen to be canonized. Born in Lombardy, Italy, the thirteenth child of a farmer, Maria Cabrini trained to be a schoolteacher and established the Missionary Sisters of the Sacred Heart to teach poor children. In 1889, she came to America where she and her followers opened many orphanages, schools, and free clinics. She became an American citizen in 1909. What do you know about this revered woman and the process of becoming a saint?

1. **How did Mother Cabrini become a saint?**
2. **Of what group is Saint Frances Cabrini the patron saint?**
3. **Who was the first person born in the United States to be recognized as a saint by the Roman Catholic Church?**
4. **How many Catholics are there in America? In the world?**

ANSWERS

1. Along with her "holy character," Mother Cabrini was credited with four miracles. By Roman Catholic law, a candidate for sainthood today must be responsible for at least two miracles to qualify as "beatified" (or Blessed). Two more miracles are required for canonization. Only the Pope can grant official recognition to a person nominated for beatification or canonization.

2. In 1950, Pope Pius XII named her patron saint of emigrants. Patron saints are seen as protectors of particular groups or places.

3. Elizabeth Ann Seton, canonized in 1975. She founded what is now often called Sisters of Charity, the first Catholic religious community that originated in the United States.

4. In 2007, there were 64.4 million Roman Catholics in the U.S. (22 percent of the U.S. population); worldwide, there were more than 1.15 billion Roman Catholics in 2005, according to the Georgetown University's Center for Applied Research in the Apostolate (CARA). Catholicism is the largest Christian church in the world.

DON'T KNOW MUCH ABOUT *Howard Hughes*

THE NAME "SPRUCE GOOSE" has been largely forgotten when compared to other aviation milestones such as the "Spirit of St. Louis" or Kitty Hawk. But in 1947, this eight-engine flying boat, largely made of birch, with room for 700 passengers, made its maiden voyage. It was piloted by its designer, Howard Hughes Jr., an eccentric genius who later became one of the world's richest men. Hughes (1905–1976) owned the Hughes Tool Company, an oil field equipment firm begun by his father. With his wealth, he gained fame as an aviator, created Trans World Airways and became a Hollywood producer. But in the mid-1950s, Hughes dropped from sight, becoming a mysterious recluse who lived in a Las Vegas hotel and never appeared in public. What do you know about this offbeat icon, subject of Martin Scorsese's biopic *The Aviator* (2004), and his fantastic flying machine?

TRUE OR FALSE?

1. **The Spruce Goose set the record for the largest wingspan of any plane ever built.**
2. **Hughes was a record-holding pilot.**
3. **As the owner of RKO Pictures, he produced such films as *Scarface*.**
4. **Before he died, Hughes wrote a bestselling autobiography.**

ANSWERS

1. True. At nearly 320 feet, it is still the largest wingspan of any airplane, according to the nonprofit group aerospaceweb.org.

2. True. He set several speed records, including an around-the-world mark of 3 days, 19 hours, 14 minutes, set in 1938.

3. True. He became a Hollywood film producer, and his most successful movies included *Hell's Angels* and *The Outlaw,* which made a star of Rosalind Russell.

4. False. In 1971, writer Clifford Irving was paid some $750,000 for a manuscript he presented as Hughes's autobiography. Irving claimed he worked on it with Hughes, but Hughes denied knowing Irving, who later confessed to the hoax.

DON'T KNOW MUCH ABOUT
Jules Verne

FRENCH NOVELIST JULES VERNE was one smart guy. He envisioned airplanes, submarines, television, guided missiles, moon rockets, and space satellites—but he did it just after the American Civil War. Born on February 8, 1828, Verne studied law in Paris but chose to become a writer, and is considered one of the founding fathers of science fiction. His first adventure tale, *Five Weeks in a Balloon* (1863), a story of crossing Africa by balloon, brought him immediate success. Verne then took readers all over the earth, under it, and above it in a number of books that also became popular films. Jules Verne died in 1905, long before many of his visions became reality. What do you know about the work of this prophet who imagined the twentieth century?

1. **In *From the Earth to the Moon* (1865), Verne's space travelers are shot out of a cannon located in what place?**
2. **Which Verne tale is about a mad sea captain who cruises beneath the oceans in a fantastic submarine?**
3. **What does Verne character Phileas Fogg attempt to do in order to win a bet?**
4. **In which novel does an expedition to a volcano lead three men on an unexpected adventure?**

ANSWERS

1. Florida, anticipating the future location of America's space launches. Verne also had these nineteenth-century astronauts splash down in the ocean.

2. *Twenty Thousand Leagues Under the Sea* (1870) tells about Captain Nemo who commanded the *Nautilus*. Nemo reappeared in *Mysterious Island* (1875).

3. In *Around the World in Eighty Days* (1873), Fogg attempts to travel around the earth in the seemingly impossible time of eighty days, an idea based on an actual trip by an American, George Train.

4. A *Journey to the Center of the Earth* (1864) takes its three heroes into the dark hollow of the earth.

DON'T KNOW MUCH ABOUT *Charles Lindbergh*

On May 20–21, 1927, American aviator Charles Lindbergh became an American icon when he made the first solo nonstop flight across the Atlantic Ocean. The press named him "Lucky Lindy" and the shy farmboy was idolized around the world. But his life of fame was shattered when his son was kidnapped in what was called the "Crime of the Century." Then his political views during World War II cost him deeply with the American public. Lindbergh died of cancer on August 26, 1974, in his home on the Hawaiian island of Maui. What do you know about the man who flew the *Spirit of St. Louis* into history?

True or False?

1. **Lindbergh was the first pilot to attempt to cross the Atlantic.**
2. **Lindbergh's Atlantic crossing was part of an Army test flight.**
3. **The famous flight took about twenty-four hours.**
4. **Lindbergh also helped start America's space program.**
5. **The Lindbergh kidnapping led Congress to pass the "Lindbergh Law."**
6. **In 1938, Nazi leader Hermann Göring presented Lindbergh with a German medal of honor.**

ANSWERS

1. False. Other pilots had crossed the Atlantic before him, but Lindbergh was the first person to do it alone nonstop.

2. False. It was a publicity stunt. In 1919, a New York City hotel owner offered $25,000 to the first aviator to fly nonstop from New York to Paris. Several pilots were killed or injured making the attempt.

3. False. Lindbergh took off in the *Spirit of St. Louis* from Roosevelt Field, near New York City, and landed near Paris. He had flown more than 3,600 miles in 33½ hours.

4. True. Lindbergh learned about Robert H. Goddard, a pioneer in rocket research, and helped obtain financial support for Goddard's experiments, which led to the development of space travel.

5. True. In 1932, the Lindberghs' infant son was kidnapped and murdered. Bruno Richard Hauptmann was convicted of the crime and executed in 1936. The "Lindbergh Law" made kidnapping a federal offense if the victim is taken across state lines or if the mail service is used for ransom demands.

6. True. Lindbergh's acceptance of the medal caused an outcry in the United States. In 1941, he joined the America First Committee, an organization that opposed American entry into World War II. He criticized President Roosevelt's policies and charged that British and Jewish groups were leading America into war.

DON'T KNOW MUCH ABOUT
Fidel Castro

HE OUTLASTED EVERY president from Eisenhower to Clinton and nearly George W. Bush. Despite economic embargos, invasions, assassination plots, counterrevolutions, and the fall of his Soviet supporters, Cuban dictator Fidel Castro remained in place until 2008. Born on August 13, 1926, Fidel Castro was a thorn in the side of American administrations since he took power in January 1959. No, he didn't always wear army fatigues, but what else do you know about the man with the famous beard and cigar who passed power to his brother Raul?

TRUE OR FALSE?

1. **Castro was the son of a Cuban army officer.**
2. **Castro was trained as a lawyer.**
3. **In 1959, Castro overthrew a corrupt military dictatorship.**
4. **The United States has never recognized Castro's government.**
5. **Cuba has one of the best human rights records in the Western Hemisphere.**

ANSWERS

1. False. Castro's father, originally an immigrant laborer, had become the wealthy owner of a large sugar plantation.

2. True. He attended Jesuit schools while growing up and went to the University of Havana's law school. He went into practice in 1950, but had been involved in rebellions in nearby Dominica.

3. True. General Fulgencio Batista had seized power in Cuba in a 1952 coup. His regime was supported by American organized crime and American sugar and fruit companies.

4. False. The United States recognized the Castro government on January 7, 1959, a few days after Castro seized power. The United States broke diplomatic relations in 1960 when Castro's government seized nearly all U.S.-owned property in Cuba and began to buy oil from the Soviet Union.

5. False. According to a 2007 report by Human Rights Watch, Cuba is the "one country in Latin America that represses nearly all forms of political dissent." Human rights groups are banned from visiting Cuba.

DON'T KNOW MUCH ABOUT *Famous Fathers*

WITH A SON of a president sitting in the White House on Father's Day 2008, it is worth remembering that "the apple doesn't fall far from the tree." Throughout American history, that has sometimes been the case. Can you figure out the famous fathers and sons in this quick quiz?

1. **Great-Grandpa was President, Grandpa was a Senator, and so was Dad. I was a Governor until January 2007.**
2. **Father ran the Securities and Exchange Commission. One of my brothers was President, another was a Senator. I'm a Senator, too.**
3. **My brother and I are historians. (I won the Pulitzer Prize!) Father was a diplomat. Grandfather was President. So was Great-Grandfather. (That should make it easy!)**
4. **My father was President, but I only got as far as Secretary of War and Ambassador to Great Britain. I also ran the Pullman Car Company, one of America's largest corporations at the time.**
5. **My father and I both went to West Point. He became President; I was a soldier, diplomat, and historian.**

ANSWERS

1. Ohio Governor Robert Taft II. He is the great-grandson of the twenty-seventh President William Howard Taft; grandson of Senator Robert Alphonso Taft I; and son of Senator Robert A. Taft, known as "Mr. Republican," who sought the nomination for President three times (1940, 1948, 1952).

2. Massachusetts Senator Ted Kennedy. His father, Joe Kennedy, was the first head of the SEC, his brother John was President, and his other brother Robert was Senator from New York.

3. Henry Adams, author of *The Education of Henry Adams,* whose brother, Brooks, was also a noted historian. Their father was Charles Francis Adams, Ambassador to Great Britain during the Civil War. Their grandfather was John Quincy Adams and great-grandfather was second President John Adams.

4. Robert Todd Lincoln, the only surviving son of Abraham Lincoln. Lincoln's other sons died before adulthood.

5. John Eisenhower, son of President Dwight Eisenhower. He served in Korea, was Ambassador to Belgium, and has written several books of military history.

DON'T KNOW MUCH ABOUT *Ansel Adams*

HE WAS AMERICA'S POET in black and white. On February 20, 1902, Ansel Adams was born in San Francisco. At age 14, he took his first photos in Yosemite National Park and he is best known for his dramatic pictures of the sweeping vistas of mountains, forests, and rivers in America's national parks. A lifelong conservationist, Adams joined the Sierra Club's Board of Directors in 1934 and was instrumental in creating new national parks, preserving existing parks, and protecting Alaska's wilds. He died in 1984. What do you know about this great American artist? Focus on this quick quiz.

1. **During an official portrait sitting, which American President did Adams lecture on preserving the Alaskan Wilderness?**
2. **During World War II, Adams photographed Japanese Americans in what notable location?**
3. **Adams helped found the Department of Photography at what famous museum?**
4. **In 1946, Adams set up something that was the first of its kind at an American college. What was it?**
5. **Where is the Ansel Adams Wilderness?**

ANSWERS

1. Jimmy Carter. Adams took Carter's official portrait in 1979. Soon after, Carter signed a bill protecting the Alaskan wilderness.

2. At Manzanar, the California internment camp to which Japanese Americans had been forcibly removed. Adams hoped to show that they were loyal, productive citizens.

3. New York City's Museum of Modern Art in 1940.

4. A photography department at the California School of Fine Arts, now the San Francisco Art Institute.

5. South of Yosemite National Park, this 229,000-acre wilderness connects the park to the John Muir Wilderness. An 11,760-foot mountain in Yosemite was also named in honor of Adams.

DON'T KNOW MUCH ABOUT *Daniel Boone*

ON JUNE 7, 1769, frontiersman Daniel Boone got his first look at the forests of present-day Kentucky, a date now celebrated as Boone Day in the Bluegrass State. One of America's first heroes, Daniel Boone (1734–1820) was a larger than life legend who deserved his fame. Daniel Boone journeyed through the rugged wilderness of the Appalachian Mountains and helped open up Kentucky. Later, he led settlers into what is now Missouri. During that journey, someone asked Boone why he was leaving Kentucky and his famous reply was, "Too crowded! I want more elbow room." What do you know about the man who once carved into a tree that he had "cilled a bar." Blaze a path through this quiz.

TRUE OR FALSE?

1. **Boone was born in North Carolina.**
2. **Boone helped blaze the Santa Fe Trail.**
3. **Boone was the model for the hero in James Fenimore Cooper's famous novels of early America.**
4. **Boone was once captured and then adopted by Indians.**
5. **Daniel Boone was famous for his coonskin cap.**
6. **This trailblazer who opened up so much territory died without any land.**

ANSWERS

1. False. He was born on November 2, 1734, in a log cabin near present-day Reading, Pennsylvania. His parents were hard-working Quakers and he was one of eleven children.

2. False. Boone helped buy a huge tract of land from the Cherokee Indians in 1775. With thirty men, he then improved the Indian trails and buffalo paths, and the resulting trail became known as the Wilderness Road, the primary route west.

3. True. Natty Bumppo, the memorable hero of *The Leatherstocking Tales* including *The Deerslayer* (1841), was modeled on Boone.

4. True. In January 1778, Boone was taken by a band of Shawnee Indians. Chief Blackfish adopted him into the tribe and gave Boone the name Shel-tow-ee (Big Turtle). The Indians plucked all of the hair from Boone's head, except for a tuft called a scalp lock, then took him to a river to "wash away his white blood."

5. False. You're probably thinking of Davy Crockett. Boone typically wore a fringed hunting shirt, deerskin leggings, and moccasins. His hair was long, tied in a pigtail, and topped with a black felt hat.

6. True. Boone sold off his Kentucky holdings to pay debts. In 1814, Congress gave him 850 acres for his services in opening the West, but Boone again had to sell the land in order to pay off his debts.

DON'T KNOW MUCH ABOUT
Historic Black Mayors

WHEN THE VOTING RIGHTS ACT of 1965 led to the removal of voting restrictions in most places, African Americans were able to help elect greater numbers of blacks to office. In 1967, the first blacks won office as mayors of large American cities. What do you know about this and other landmarks in African-American election history?

1. **Who was the first black mayor of a major American city?**
2. **Also elected in 1967, but inaugurated slightly later, Richard Hatcher became the second black mayor of what large American city?**
3. **Which sharecropper's son led a large California city for twenty years?**
4. **In 1973, Maynard Jackson became the first black mayor of what major Southern city?**
5. **Who defeated Rudy Giuliani to become the first African American mayor of New York City?**
6. **Which grandson of slaves became the first elected black governor in the United States?**

ANSWERS

1. Carl B. Stokes (1927–1996) served as mayor of Cleveland from 1967 to 1971. The first black to be elected to head a major American city, he was reelected in 1969. Stokes later worked as a newscaster in New York City, served as a judge on the Cleveland Municipal Court, and was named the U.S. ambassador to Seychelles from 1994 to 1996.

2. Hatcher served as mayor of Gary, Indiana, from 1968 to 1987. Hatcher won reelection four times, but was defeated in a bid for another term in the 1987 Democratic primary.

3. Thomas Bradley was the first African-American mayor of Los Angeles. In 1973, he became leader of what was then the third largest U.S. city and was reelected four more times—to date still the only person ever elected mayor of Los Angeles more than three times.

4. Atlanta, Georgia.

5. David Dinkins defeated Giuliani in the 1989 election. In 1993, Dinkins ran for reelection, again against Giuliani, but was defeated and left office in 1994.

6. Douglas Wilder was the chief executive of Virginia from 1990 to 1994.

DON'T KNOW MUCH ABOUT *Norman Rockwell*

A DROPOUT WHO left school at sixteen to study art, he became America's most beloved painter. Born in New York, Norman Rockwell (1894–1978) was still a teenager when he became art director of *Boy's Life*, the Boy Scout magazine. Then, in May 1916, he began the career for which he is best known when he painted his first cover for America's favorite magazine, *The Saturday Evening Post.* During the next half century, Rockwell painted more than 300 covers for the *Post.* Whimsical, heartwarming, celebrating an idyllic America, Rockwell's work mirrored America in the twentieth century. But he could also depict the dark side of the American dream. What do you know about "America's Artist"?

1. **What wartime speech by President Roosevelt inspired a series of famous Rockwell paintings?**
2. **In a memorable May 1943 *Post* illustration, what wartime character did Rockwell bring to life?**
3. **In "The Problem We All Live With," his first cover for *Look* magazine in 1964, what notable social issue did Rockwell depict?**
4. **What headline-making crime did a 1965 *Look* cover illustrate?**
5. **Where is the Norman Rockwell Museum?**

ANSWERS

1. "The Four Freedoms" in 1943: *Freedom of Speech, Freedom to Worship, Freedom from Want,* and *Freedom from Fear.*

2. "Rosie the Riveter." The brawny woman in coveralls munching a sandwich with a rivet gun in her lap was a tribute to the millions of women who worked in the defense industry.

3. Desegregation. The painting showed a small black girl being escorted to school by federal marshals.

4. *Murder in Mississippi* depicted the killings of three civil rights workers in 1964.

5. Stockbridge, Massachusetts, Rockwell's home from 1953 to the time of his death in 1978.

DON'T KNOW MUCH ABOUT *Nostradamus*

IN THE WAKE of the September 11, 2001 terror attacks, the name Nostradamus was once again in the news. Suddenly, a sixteenth-century astrologer was one of the most searched names on the Internet, because he had supposedly predicted the attacks on the World Trade Center. Nostradamus (pronounced nahs-truh-DAY-muhs) was the Latinized name of Michel de Notredame, a physician born in 1503 in southern France. Nostradamus won lasting fame when King Henry II of France died as predicted in his book *Centuries*, a series of poetic prophecies written in 1555. Are you skeptical this mystic foretold world events? Check your crystal ball with this quick quiz.

TRUE OR FALSE?

1. **During World War II, the Nazis issued a version of Nostradamus's prophecies to convince Germans of the Nazis' ultimate victory.**
2. **Nostradamus was the son of the prominent Archbishop of Notre Dame.**
3. **During the great plague, Nostradamus was a physician who saved many lives.**
4. **Nostradamus predicted the end of the world in the year 2550, a thousand years after his book appeared.**

Answers

1. True. The Nazis also believed that the prophecies would convince their enemies of their inevitable victory.

2. False. He was born into a Jewish family which converted to Roman Catholicism.

3. True. A well-educated physician, Nostradamus embraced remedies that were said to be more helpful than the popular bleeding techniques of that time.

4. False. He told his son that he made predictions up to the year 3797, when the world would end or be transformed.

DON'T KNOW MUCH ABOUT *The Wright Brothers*

ON DECEMBER 17, 1903, self-taught engineer-inventors Wilbur and Orville Wright did something that other people had only dreamed of for centuries. In a 750-pound plane powered by a 12-horsepower motor and launched from the sand dunes of Kitty Hawk, North Carolina, they flew the first heavier-than-air craft. Surprisingly, their initial success didn't cause much of a stir. Even the U.S. Army was dubious and refused to offer the Wrights a contract for more than three years. But their first flights launched a revolutionary era of inventiveness and aviation heroics. It is astonishing to think that those brief flights would lead to a moon landing only sixty-six years later, while also creating a huge industry in wheeled luggage and small bags of roasted nuts. What else do you know about the Wright Brothers? Take off with this quick quiz.

1. **What businesses did the Wright Brothers run in Dayton, Ohio?**
2. **How long was the Wright Brothers first flight?**
3. **Before the brothers took flight, Orville developed a crucial piece of technology in 1901. What did he design?**
4. **In September 1908, Orville Wright took the first passengers on a flight. What unfortunate distinction does one passenger, Thomas Selfridge, hold?**

ANSWERS

1. Although best known as bicycle shop owners, the Wright Brothers were also printers and publishers of a small newspaper, Dayton's *West Side News.*

2. Twelve seconds. On later attempts that day, older brother Wilbur stayed aloft for 59 seconds and traveled 852 feet.

3. Orville developed the first wind tunnel, crucial in developing the airplane.

4. Selfridge was the first man to die in a plane crash of injuries suffered on September 17, 1908.

DON'T KNOW MUCH ABOUT *Winston Churchill*

BORN ON NOVEMBER 30, 1874, Winston Churchill was a child of the nineteenth century who became one of the greatest men of the twentieth century. The son of a British lord and an American heiress, he had a long career in politics before he became Britain's wartime Prime Minister on May 10, 1940, promising "blood, toils, tears and sweat." His pugnacious style, rousing speeches, and signature "V for Victory" sign helped rally Great Britain through the darkest days of the fight against Hitler. What else do you know about Churchill, who died in 1965?

1. **During what war was Churchill captured as a correspondent?**
2. **Churchill harshly criticized Prime Minister Neville Chamberlain over what key historical event?**
3. **In a radio broadcast, what country did he call "a riddle wrapped in a mystery inside an enigma"?**
4. **"We shall fight in the fields and in the street, we shall fight in the hills; we shall never surrender." This famed speech followed what crucial event?**
5. **In a famous 1946 speech delivered in Missouri, how did Churchill describe Eastern Europe?**

ANSWERS

1. Churchill was captured in South Africa during the Boer War in 1899. His escape made him a national hero and he was elected to Britain's Parliament in 1900.
2. Chamberlain's Munich agreement of 1938 with Hitler, widely viewed as an act of "appeasement" that led to Germany's attempt to conquer Europe.
3. It was his description of Russia in 1939.
4. The evacuation of British troops from Dunkirk, France, 1940.
5. He said an "Iron Curtain" had fallen across the continent. Though not the first time the phrase was used, it was certainly the most famous and lasting use.

DON'T KNOW MUCH ABOUT
Mark Twain

AMERICA DOESN'T HAVE a national holiday to honor a writer. But if we did, maybe it should be one devoted to Samuel Langhorne Clemens, born in Missouri on November 30, 1835. Best known by his pen name, and often viewed as the creator of such children's classics as *A Connecticut Yankee in King Arthur's Court, The Prince and the Pauper,* and *The Adventures of Tom Sawyer,* Mark Twain was much more. In a distinctly American style, Twain wrote biting satire that poked fun at America's manners and corrupt politics. *The Adventures of Huckleberry Finn* (1884), his masterwork, is now a controversial classic. But Twain would surely remind people that he once said that a classic is "a book which people praise and never read." Although he famously told a newspaper in 1897, "The report of my death was an exaggeration," Twain, in fact, died in 1910. What else do you know about one of America's greatest writers? Take this quick quiz.

1. **Where did he get his pen name "Mark Twain"?**
2. **How did Twain serve during the Civil War?**
3. **What short story gave Twain his national fame?**
4. **Which famous general's autobiography did Twain publish?**

ANSWERS

1. As a steamboat pilot on the Mississippi River, he knew this phrase meant that the water is two fathoms (12 feet) deep.

2. In 1861, Clemens joined a group of irregular Confederate cavalry from Missouri, deserting after a few weeks' time. The experience served as the source of a short memoir, "The Private History of a Campaign That Failed."

3. "The Notorious Jumping Frog of Calaveras County" (1865), based on a tale Twain heard while working in the California gold fields, was a national sensation.

4. His firm, Charles L. Webster, published Ulysses S. Grant's *Memoirs*, a critical and commercial success.

DON'T KNOW MUCH ABOUT *Theodore Roosevelt*

WHEN A MAN in a crowd once asked if Teddy Roosevelt had been drinking, the reply came, "Oh, no. He needs no whiskey to make him feel that way—he intoxicates himself by his own enthusiasm." Born on October 27, 1858, Theodore Roosevelt was elected William McKinley's Vice President on November 7, 1899. When McKinley was assassinated in September 1901, "Teddy"—a nickname he detested—became America's youngest President at age 42. Overshadowed as a President by his cousin, Franklin Delano, TR has risen in the estimation of historians. What do you know about the man who inspired the Teddy Bear and the famous palindrome "A Man, a Plan, a Canal—Panama"?

1. **During the Spanish-American war, Roosevelt joined the First Volunteer Cavalry, a group best known by what name?**

2. **Roosevelt was Governor of New York when nominated for Vice President. What other notable post had he held in New York?**

3. **Among many accomplishments, Roosevelt achieved an honor that only two other Presidents have won. What is it?**

4. **What was his most famous pronouncement, actually an African proverb?**

5. **After winning a second term as a Republican President in 1904, Teddy chose not to run in 1908. When he changed his mind for the 1912 elections, what was his party?**

ANSWERS

1. The admiring press christened them "Rough Riders." For leading an assault on Kettle Hill near San Juan, Cuba, Roosevelt became an instant war hero.

2. He was New York City Police Commissioner, where he earned a reputation as a corruption fighter and reformer.

3. Roosevelt won the Nobel Peace Prize in 1906 for helping to end the Russo-Japanese War. Woodrow Wilson won the Nobel Peace Prize in 1919, and Jimmy Carter in 2002.

4. "Speak softly and carry a big stick; you will go far." Roosevelt rarely spoke softly.

5. The Bull Moose, or Progressive, Party. Democrat Woodrow Wilson won the election, defeating the incumbent Republican, William Taft, Roosevelt's Secretary of War.

DON'T KNOW MUCH ABOUT *Dr. Seuss*

IF YOUR BOOK was turned down by more than forty publishers, "What would you do?" If you were Theodor S. Geisel, get a friend to publish the book. Thus was born Dr. Seuss. Actually born on March 2, 1903, Theodore Seuss Geisel first turned his knack for words and pictures to advertising and editorial cartoons. But Dr. Seuss, who died in 1991, influenced entire generations of children with his nonsensical poems that put "See Spot Run" on the endangered species list. So what do you know about Seuss? Heaven save us / Try this quick quiz.

1. **Inspired by the rhythmic sound of an ocean liner's engine, what was Seuss's first book?**
2. **Which Seuss classic used just 220 words?**
3. **Boris Karloff once made his voice rather scary / But in a remake, he was played by Jim Carrey. Who is he?**
4. **Here's a clue that may surprise you / What did Seuss do / During the War known as Two?**

ANSWERS

1. *And to Think That I Saw It on Mulberry Street.* The idea came to Seuss on an ocean cruise.

2. *The Cat in the Hat,* written in response to the 1954 reports of poor reading in America.

3. The Grinch.

4. He drew anti-Nazi and anti-Japanese propaganda cartoons, images sharply at odds with his whimsical children's drawings.

DON'T KNOW MUCH ABOUT *Pope John Paul II*

IN CELEBRATION OF the millennium year 2000, Pope John Paul II made a pilgrimage to the Holy Land, with visits to Jerusalem, Bethlehem, and other biblical sites. During his pontificate, which began in 1978, the Polish-born John Paul II (originally Karol Wojtyla) was the most traveled leader of the Roman Catholic church in papal history. He died on April 2, 2005, at age 84. What else do you know about the late popular leader and his office?

1. **Who was the first Pope?**
2. **What distinction did John Paul II hold that no Pope had for 450 years?**
3. **In 1984, the U.S. restored diplomatic relations with the Vatican. How long ago were these diplomatic relations with the Vatican banned by Congress?**
4. **Besides leading the Roman Catholic church, the Pope is head of what state?**
5. **Which biblical site in Iraq did John Paul II visit while Saddam Hussein was in power?**

ANSWERS

1. St. Peter, the first bishop of Rome, is considered the first Pope.
2. He was the first non-Italian Pope in that time.
3. The ban on diplomatic relations with the Vatican dates to 1867.
4. The Pope is Sovereign of the State of Vatican City, which was created as an independent state in 1929.
5. The city of Ur, the traditional birthplace of the Patriarch Abraham, although some biblical scholars dispute this.

DON'T KNOW MUCH ABOUT *Martin Luther King Jr.*

ON JANUARY 15, 1929, the U.S. Senate ratified the Kellogg-Briand Pact, a treaty designed to peacefully settle international disputes. For his role in creating that pact, Secretary of State Frank Kellogg won the Nobel Peace Prize. On the very same day, future Nobel Peace Prize winner, Martin Luther King Jr., was born in Atlanta, Georgia. After receiving his doctorate from Boston University, King returned to the South to lead the nonviolent civil rights movement, changing the face of America. His death in April 1968 remains controversial to this day, as allegations of a conspiracy continue to swirl around his assassination. In 1986, King's birthday was first observed as a national holiday, to be celebrated on the third Monday in January. What else do you know about this Baptist minister who was one of the most profoundly influential Americans of the century?

TRUE OR FALSE?

1. King's first prominent protest action was a sit-in at a Woolworth's segregated lunch counter.

2. In 1957, King founded the National Association for the Advancement of Colored People (NAACP) to battle racism and segregation.

3. King's most famous address, the "I Have a Dream" speech, was delivered to a mass demonstration in Washington, D.C.

4. King was the last American to receive the Nobel Peace Prize, which he won in 1964.

ANSWERS

1. False. He led the 1955 Montgomery, Alabama, bus boycott, following the arrest of Rosa Parks.

2. False. In 1957, King founded the Southern Christian Leadership Conference (SCLC). The NAACP was begun in 1909.

3. True. More than 250,000 people heard this speech at the August 1963 Civil Rights March on Washington.

4. False. Since 1964, six Americans have won the Peace Prize. Agricultural-scientist Norman Borlaug (1970), Secretary of State Henry Kissinger (1973), Hungarian-born writer Elie Wiesel (1986), Jody Williams of the International Campaign to Ban Land Mines (1997), President Jimmy Carter (2002), and Vice President Al Gore (2007).

DON'T KNOW MUCH ABOUT *Young Martin Luther King Jr.*

MARTIN LUTHER KING DAY honors the martyred civil rights leader on the third Monday in January. The campaign to establish a holiday honoring King began soon after he was assassinated in 1968. King's career as a civil rights leader is well known. But what do you know about his early life?

TRUE OR FALSE?

1. **King was born in Montgomery, Alabama.**
2. **His father was pastor of the Ebenezer Baptist Church in Atlanta.**
3. **King was a high school dropout.**
4. **He met his future wife, Coretta, while studying in Boston.**
5. **King's first church was his father's church in Atlanta.**
6. **Martin Luther King Jr. was not named "Martin" at birth.**

ANSWERS

1. False. King was born in Atlanta on January 15, 1929. The second oldest child, he had an older sister, Christine, and a younger brother, A. D. The young Martin was usually called M. L.

2. True. One of Martin's grandfathers, A. D. Williams, also had been pastor there.

3. False. Martin did so well in high school that he skipped both the ninth and twelfth grades. He entered Atlanta's Morehouse College when he was only 15.

4. True. After Seminary in Pennsylvania, King went to Boston University to earn his Ph.D. He met Coretta Scott, a music student, there and they married in 1953.

5. False. In 1954, King became pastor of the Dexter Avenue Baptist Church in Montgomery, Alabama. The next year, he became a leader in the protest of Montgomery's segregated bus system following the arrest of Rosa Parks for disobeying a city law requiring that blacks surrender their bus seats when white people wanted to sit in their seats or in the same row.

6. True. His given name was Michael, and his father changed both his and his son's name to Martin, in honor of Protestant Reformer Martin Luther, following a visit to the Holy Land.

DON'T KNOW MUCH ABOUT *President George Washington*

"FIRST IN THE HEARTS of his countrymen" and first in the presidency was George Washington, who died on December 14, 1799. Once elected, Washington said, "I walk on untrodden ground," knowing that his every move would establish lasting precedents for the office. One difference when Washington was elected in 1789 was that the first President's first inauguration took place on April 30, 1789, not in January, as is now required by the Constitution. What else do you know about the first President's presidency? And remember, do not tell a lie.

TRUE OR FALSE?

1. **Since Washington, D.C., didn't exist in 1789, the first President took the oath of office in New York City.**
2. **In the first election, Washington swept the electoral votes of all thirteen states.**
3. **Washington's running mate was John Adams.**
4. **Washington's first Cabinet included Alexander Hamilton (Treasury), Thomas Jefferson (Secretary of State), Benjamin Franklin (Education), and John Paul Jones (Defense).**
5. **The last time George Washington led troops into the field was as President.**

Answers

1. True. Washington was inaugurated at Federal Hall in New York City, the nation's first capital. The capital was moved to Philadelphia in his second term.
2. False. Washington was elected unanimously, but carried only the ten states that actually voted. New York failed to choose its "electors" in time, and both North Carolina and Rhode Island had not yet ratified the Constitution and did not participate in the first presidential election.
3. False. There were no political parties at the time and no "presidential tickets." Adams became Vice President because he was the second-place finisher. That was how the Vice President was chosen until the Twelfth Amendment was ratified in 1804.
4. False. While the first two are accurate, there were no Education or Defense departments back then. In fact, the new Constitution did not mention a Cabinet at all. Washington's first group of department heads was rounded out by War Secretary Henry Knox, a veteran of the Revolution for whom Fort Knox is named, and Attorney General Edmund Jennings Randolph, a prominent Virginian.
5. True. The Whiskey Rebellion was a revolt against a federal tax passed on corn whiskey, which was how most farmers used their surplus corn. Washington led 13,000 militiamen into the Pennsylvania wilderness to put down this anti-tax revolt, a force larger than any he had commanded during the Revolution.

DON'T KNOW MUCH ABOUT *"Dead Presidents"*

INDEPENDENCE DAY USUALLY reminds us of America's birth and George M. Cohan's "Yankee Doodle Dandy," who was "Born on the Fourth of July." But in one of the strange ironies of history, three prominent American Presidents died on the Fourth of July. And a fourth President was born on the Fourth. Which Presidents are connected to the glorious Fourth? Try this quick quiz.

1. **A Governor, Secretary of State, and Vice President, his epitaph includes "Father of the University of Virginia."**
2. **Like George Bush, he was a former Vice President who served a single term and had a politically famous son.**
3. **A veteran of the Revolution, he was a Senator, Vice President, and Minister to France and England, and his Presidency was called the "Era of Good Feelings."**
4. **A former Governor and Vice President, he was sworn into office by his father, a Justice of the Peace, when his predecessor died.**

Answers

1. The third President, Thomas Jefferson, died a little after noon at Monticello on July 4, 1826, the fiftieth anniversary of the Declaration of Independence which, of course, he wrote.

2. The second President, John Adams died on the same day as Jefferson, about six hours later in Quincy, Massachusetts. His son, John Quincy Adams, became the sixth President, making them the first father and son American Presidents.

3. The fifth President, James Monroe died on July 4, 1831 in New York City where he moved after financial problems forced him to sell his Virginia home.

4. The thirtieth President, Calvin Coolidge was born in Plymouth, Vermont, on July 4, 1872, and took office when Warren G. Harding died.

DON'T KNOW MUCH ABOUT *Famous Spies*

THE RELEASE IN 2003 of fifty-year-old transcripts of Senator Joseph McCarthy's secret hearings about Communists in America was a stark reminder of that fearful era. So is the still controversial event: the execution of Ethel and Julius Rosenberg on June 19, 1953. Electrocuted at Sing Sing Prison, they became the first civilians to suffer the death penalty in an espionage trial. What do you know about the Rosenbergs and some other notorious spy cases? Investigate in this quick quiz.

1. **What were the Rosenbergs accused of?**
2. **Who was David Greenglass?**
3. **What did the Walker family do?**
4. **Who was Aldrich Ames?**
5. **What did Robert Hanssen do?**

ANSWERS

1. Charged with passing secrets about the atomic bomb to the Soviets, they pleaded innocent and then pleaded the Fifth when asked if they were Communists. For years, many people believed that the Rosenbergs did not get a fair trial and that their sentence was too harsh, especially as they had two small boys. The case was appealed to the Supreme Court of the United States, but all appeals were denied and President Eisenhower rejected two pleas for clemency.

2. The brother of Ethel Rosenberg, Greenglass had worked as a machinist at Los Alamos, on the project to make an atomic bomb. The government charged that he supplied the information used to build the first Soviet atomic bomb and he implicated the Rosenbergs. Sentenced to fifteen years in prison, Greenglass later admitted he had lied about his sister to save himself. Recently opened Soviet files confirm that Julius was a spy but had not passed on atomic secrets, and that Ethel knew about this but was not involved.

3. Navy communications specialist John Walker, his son Michael, and his brother Arthur sold secrets to the Soviets for decades. All received lengthy prison terms.

4. A CIA operative, Ames also spied for the Soviets. In spite of a modest salary, he owned an expensive home and cars bought with his espionage profits, but went undetected for nine years. His deceit may have cost the lives of hundreds of agents working for the United States. He is serving a life sentence.

5. An FBI counterintelligence expert, Hanssen sold secrets to the Soviets for more than twenty years. He also is serving a life sentence.

DON'T KNOW MUCH ABOUT *White House Women*

CAN A WOMAN win the White House? Will a future presidential election bring new meaning to the words "First Lady"? The candidacy of Senator Hillary Clinton made those questions less abstract. But it is clear that when it comes to politics, women have come a long way. How much do you know about women's achievements in politics?

TRUE OR FALSE?

1. **Elizabeth Dole was the first female Cabinet member.**
2. **Bella Abzug was the first woman in Congress.**
3. **Women could not vote until the Nineteenth Amendment passed in 1920.**
4. **The first woman on a major party's presidential ticket was Geraldine Ferraro.**
5. **The first woman in the White House was Abigail Adams.**

ANSWERS

1. False. Frances Perkins was Franklin D. Roosevelt's Labor Secretary in 1933.

2. False. Jeanette Rankin was the first woman elected to Congress from Montana in 1916.

3. False. Women could vote in several state and local elections before then.

4. True. Geraldine Ferraro was Walter Mondale's Democratic running mate in 1984.

5. True. John and Abigail Adams moved into the unfinished presidential palace in June 1800.

DON'T KNOW MUCH ABOUT *First Ladies*

THROUGHOUT HISTORY, AMERICA'S First Ladies have been both adored and scorned. Some have avoided the limelight; others have shone brilliantly in the role. What do you know about these extraordinary women who stand beside the President?

1. **Which First Lady said, "I live a dull life here. . . . I am more like a state prisoner than anything else"?**
2. **After her husband suffered a stroke, which First Lady practically acted as president?**
3. **Which First Lady became America's delegate to the General Assembly of the United Nations?**
4. **Which First Lady had a career as an "inquiring photographer" for the *Washington Post* before her marriage?**
5. **Who was the First Lady elected to public office?**
6. **Which presidential nominee's wife later campaigned to become President herself?**
7. **So far, only two First Ladies have also been "First Moms." Who are they? And what else do these two often outspoken ladies have in common?**

ANSWERS

1. Martha Washington, on life as first lady in New York City.
2. Woodrow Wilson's second wife, Edith Bolling Wilson.
3. Eleanor Roosevelt, from 1945 to 1951.
4. Jackie Lee Bouvier, who married John F. Kennedy in 1953.
5. Senator Hillary Rodham Clinton.
6. Elizabeth Dole, wife of Senator Bob Dole, who was elected U.S. Senator from North Carolina in 2002.
7. Abigail Adams, wife of second President John Adams and mother of sixth President John Quincy Adams; Barbara Bush, wife of forty-first President George H. W. Bush and mother of forty-third President George W. Bush. Both were married at age 19.

DON'T KNOW MUCH ABOUT *Senator Joseph McCarthy*

"McCarthyism." It is a word that still inspires strong feelings more than fifty years later. On December 2, 1954, the U.S. Senate voted 65 to 22 to censure and condemn the man who inspired it, Wisconsin Senator Joseph McCarthy. McCarthy gained worldwide attention in 1950 by charging that Communists had infiltrated the government. Some praised him as a patriot; others condemned him for publicly accusing people of disloyalty without sufficient evidence. What do you know about the man and his times, which inspired George Clooney's 2005 movie, *Good Night and Good Luck*?

1. **What was McCarthy's famous nickname?**
2. **In what notable first was McCarthy involved?**
3. **Who coined the word "McCarthyism"?**
4. **What bestseller includes a staunch defense of Senator McCarthy?**
5. **What became of McCarthy?**

ANSWERS

1. "Tailgunner Joe." As an intelligence officer stationed in World War II, he participated in combat bombing missions, although he was not wounded in action as he later claimed.

2. The "Army-McCarthy" hearings he held were the first nationally televised Senate hearings.

3. The word was first used by the cartoonist Herblock in the *Washington Post*. The word was a synonym for baseless mudslinging, and in the cartoon, "McCarthyism" was crudely lettered on a barrel of mud. But McCarthy proudly used the word himself as a book title, *McCarthyism: The Fight for America*.

4. *Treason* by Ann Coulter.

5. He remained in the Senate, but was largely ignored by Congress, the White House, and the media. The alcoholic McCarthy's drinking eventually caused a liver ailment and he died of acute hepatitis on May 2, 1957.

EXCEPTIONAL PLACES

DON'T KNOW MUCH ABOUT *Brazil*

IT IS LARGE. It is populous. It is east of the United States. And more than four hundred years ago, it was named after a tree. On April 22, 1500, when a Portuguese fleet landed on the coast, Portugal laid claim to Brazil. They found trees there with the color of a glowing ember—called *brasa* in Portuguese—and called them "brazilwoods." In other words, the country was named for the trees. By many measures, Brazil is one of the world's most significant countries. Yet few Americans know much more than "The Girl from Ipanema" when it comes to the world's fifth largest country. Take this quick quiz.

TRUE OR FALSE?

1. **Brazil is South America's largest country in both area and population.**
2. **The Amazon rain forest is the world's second largest tropical rain forest.**
3. **The Amazon River is the world's second longest river.**
4. **Brazil has more Catholics than any other country.**
5. **Only Colombia produces more coffee than Brazil.**
6. **The name Amazon is a native Indian term.**
7. **Brazil abolished slavery before America did.**

ANSWERS

1. True. In 2007, Brazil ranked as the world's fifth largest nation in both area and population. Only China, India, the United States, and Indonesia have more people. About half the people of South America live in Brazil, more people than all the other South American nations combined. São Paulo, with about 11 million people, ranks as one of the world's largest cities and Rio de Janeiro has about 6 million people. Brazil occupies almost half the continent.

2. False. It is the largest, covering approximately 2 million square miles in the Amazon River Basin. About two-thirds of the rain forest lies in Brazil. The Amazon rain forest contains a wider variety of plant and animal life than any other place in the world.

3. True. The Nile River is longer, but the Amazon carries more water than any other river—more than the Mississippi, Nile, and Yangtze rivers together. At 4,000 miles, it is longer than the highway route between New York City and San Francisco.

4. True. Brazil was a Portuguese colony from 1500 to 1822 and the early colonists introduced Roman Catholicism. In 2005, Brazil had about 184 million Catholics, an estimated 74 to 79 percent of the country's population, but many of them are considered nominal Catholics, and the number is declining.

5. False. In 2004, the nation was the world's largest "coffeepot," producing 34 percent of the total annual coffee crop, according to the International Coffees Organization.

6. False. During 1541 and 1542, a Spaniard, Francisco de Orellana, led the first European exploration of the river. When attacked by what appeared to be female Indian warriors, the Spaniards called their attackers Amazons, after the female warriors in Greek mythology.

7. False. Brazil's slaves, brought from Africa to work the sugar fields, were freed in 1888.

DON'T KNOW MUCH ABOUT *Pennsylvania Avenue*

OFTEN CALLED "AMERICA'S MAIN STREET," the thoroughfare that runs through the heart of Washington, D.C., is probably best known for a single address: 1600 Pennsylvania Avenue, the White House. For more than a decade, the stretch of the avenue in front of the White House has been closed to vehicles, but pedestrians may still use the street there. What else do you know about this magnificent boulevard, scene of so much history? Stroll through this quick quiz.

1. **What event caused the closure of Pennsylvania Avenue in front of the White House?**
2. **Why is it named for Pennsylvania?**
3. **How long is Pennsylvania Avenue?**
4. **What did George Washington call the street?**
5. **What two major events took place on the street in 1865?**

ANSWERS

1. The Secret Service ordered the temporary closure following the Oklahoma City bombing in 1995. After September 11, 2001, the change was made permanent.

2. The symbolically important street was named for the Keystone State as consolation for moving the nation's capital from Philadelphia.

3. It runs for seven miles inside Washington. While the stretch from the White House to the Capitol is most familiar, it continues on the other side of the Capitol for many miles, over the Anacostia River and into Maryland. The street continues northwest past the White House ending in Georgetown.

4. Laid out by the city's chief planner Pierre L'Enfant, Pennsylvania Avenue was one of the earliest streets constructed in the federal city. After inspecting L'Enfant's plan, Washington referred to it as a "Grand Avenue." While the "grand avenue" was little more than a dirt road at the time, Thomas Jefferson planted it with rows of fast growing Lombardy poplars.

5. Abraham Lincoln's funeral followed two weeks later by a parade celebrating the end of the Civil War.

DON'T KNOW MUCH ABOUT *Nevada*

SINCE OLDEN DAYS when prospectors came looking for silver, to the boom-time era of modern Las Vegas, people have come to Nevada in search of fortunes. A member of the union since 1864, Nevada was the fastest growing state in the nation for nineteen straight years until it was nosed out by Arizona in 2006. (Nevada's growth rate was 3.5 percent while Arizona's was 3.6 percent in 2006.) With its beautiful mountains and deserts, the thirty-fifth state is more than slots and Elvis impersonators. What do you know about the "Silver State"? Ante up in this quick quiz.

TRUE OR FALSE?

1. **Once called Silver City, the state capital of Carson City was renamed in honor of local boy Johnny Carson.**
2. **Created during the Mexican War, it is known as the "Battle Born State."**
3. **Nevada is the driest state in the Union.**
4. **Prostitution is legal statewide in Nevada.**
5. **Carson City and Reno both lie west of Los Angeles.**

ANSWERS

1. False. The city was named for famed explorer and mountain man Kit Carson.
2. False. Nevada gained statehood during the Civil War when Abraham Lincoln needed more pro-Union Republican votes in the Senate—which makes the "Running Rebels" of the University of Nevada, Las Vegas, a slightly inaccurate nickname.
3. True. In terms of precipitation. Other forms of liquids are widely available.
4. False. "Brothel prostitution" is legal only in certain counties and not in such cities as Las Vegas, Reno, Carson City, and Lake Tahoe.
5. True. Los Angeles lies to the east of these Nevada cities as the southern California coast cuts back in an easterly direction.

DON'T KNOW MUCH ABOUT *Japan*

A YEAR AFTER Matthew Perry first sailed four American ships into Tokyo Bay in 1853, Japan and America signed the Kanagawa Treaty in March 1854. The beginning of formal relationships between the countries, this agreement opened up Japanese ports to foreign trade. The United States and Japan have come a long way since then, having fought each other in World War II and gone on to dominate the world economy. Did you see *The Last Samurai* and think you know about Japan?

1. **Where does the word "Japan" come from?**
2. **Who are the Ainu?**
3. **Is Japan the most densely populated country?**
4. **What is the Shinkansen?**
5. **True or False? Japan's gross domestic product is greater than America's.**
6. **Is *The Last Samurai* a true story?**

Answers

1. The Japanese call their country Nippon or Nihon, which means "source of the sun" and the Emperor was traditionally believed to be a descendant of the sun god. But the name Japan comes from "Zipangu," the Italian name given to the country by the famed European explorer Marco Polo.

2. A group of people who may have been the first inhabitants of what is now Japan. Scientists are uncertain about the ancestry of the Ainu who have lived mostly on Hokkaido, Japan's northernmost main island.

3. No. About 126 million people are crowded on its four main islands, making Japan one of the most densely populated countries in the world, with 838 people per square mile compared with only 83 in the United States. But the tiny principality of Monaco is the world's most densely populated country, according to U.N. figures from 2005.

4. High-speed electric trains, these are the bullet trains linking Tokyo with other cities at speeds of up to 186 mph.

5. False. According to the World Bank's 2006 listings, Japan's GDP of $4,340,133,000 (in U.S. dollars) ranks behind that of the United States. In the World Bank list, Japan is followed by Germany, China, and the United Kingdom. The International Monetary Fund 2006 list, however, ranks Japan third in GDP because it counts the European Union, with its current twenty-seven member states, as the world leader in GDP.

6. Although based loosely in fact, *The Last Samurai* is the fictional tale of a Civil War veteran (Tom Cruise) who travels as a mercenary to Japan in 1868. While the legendary samurai are known as strong and courageous warriors, they were in reality an elite and largely idle class that spent more time drinking and gambling than fighting gloriously.

DON'T KNOW MUCH ABOUT *Columbus*

IN FOURTEEN HUNDRED AND NINETY-TWO / *Columbus got lost on the ocean blue.* But as America celebrates each passing Columbus Day, the shining legend of the great explorer has been considerably darkened. The image of the intrepid sailor who accidentally "discovered" America has been tarnished, especially in light of his treatment of the natives he encountered and misnamed *Indians.* Still, this Italian sailor who sailed in the service of Spain is honored across America and around the world with cities, towns, and rivers named in his honor. How well do you know the places named for Columbus? Or are you just as lost as Columbus was when he thought he'd found China?

1. **Of the many American cities named for Columbus, which two are state capitals?**
2. **Where does the major river named in his honor get its start?**
3. **First called Aspinwall by Americans, which Latin American capital city honors the Spanish form of his name?**
4. **Which Asian capital is named in Columbus's honor?**

ANSWERS

1. Columbus, Ohio, and Columbia, South Carolina.
2. The Columbia River, which forms the border between Oregon and Washington, starts in British Columbia, Canada, a province also named for Columbus.
3. Colón, Panama.
4. Colombo, Sri Lanka (formerly Ceylon), was renamed by the Portuguese to honor Columbus.

DON'T KNOW MUCH ABOUT
Ohio

A LITTLE MORE than two hundred years ago, Ohio became the seventeenth state on March 1, 1803. Carved out of the Northwest Territory, a large area captured during the Revolution, Ohio has been a historically significant state. Home of two of the most famous astronauts (John Glenn and Neil Armstrong), Ohio is also called the Mother of Presidents. If you don't know why, take this quick quiz.

1. **While Virginia has produced the most Presidents (8), the Mother of Presidents is in second place. How many Presidents were Ohioans? How many can you name?**
2. **John Chapman, born in Massachusetts, was renowned for his exploits in Ohio. Who was he?**
3. **What's a buckeye?**
4. **Was Ohio's Mound City Group the first rock band in America?**
5. **Can you name Ohio's capital and largest city?**
6. **What distinction goes to Ohio's Oberlin College?**

ANSWERS

1. Seven. They are Ulysses S. Grant, Rutherford B. Hayes, James Garfield, Benjamin Harrison, William McKinley, William H. Taft, and Warren G. Harding.

2. Chapman, born in 1774, is better known as Johnny Appleseed because he supposedly spent his life traveling through Ohio, planting orchards as the settlers moved west.

3. The nickname comes from the tree whose horse chestnuts looked like the eye of a buck. Early pioneers cut down many buckeyes to build log cabins.

4. It's not a rock 'n' roll band, but a national park preserving a group of Indian burial mounds left by the Mound Builders, Indians who date to prehistoric times.

5. With more than 730,000 people, Columbus was America's fifteenth largest city in 2005. Cleveland ranked thirty-ninth.

6. It was America's first coeducational college, awarding degrees to women in 1841.

DON'T KNOW MUCH ABOUT *Central Park*

IF IT IS NOT THE WORLD'S most famous park, it certainly is near the top of the list. Famed, often filmed, and much feared for its once high crime rate, Central Park is New York City's big backyard and playground. When city leaders called for a grassy refuge more than 150 years ago, 843 acres were carved out after the usual haggling and land scandals. The park was then laid out according to the 1857 "Greensward Plan" of landscape architects Frederick Law Olmsted and Calvert Vaux. What else do you know about this Big Apple landmark?

TRUE OR FALSE?

1. **Central Park attracts some 10 million visitors every year.**
2. **There are fifty-one statues, monuments, and sculptures dotting the park.**
3. **The largest gathering in Central Park was for a Simon and Garfunkel concert.**
4. **A Hollywood legend, Central Park has "starred" in more than 170 films.**
5. **Like the subway, the famed park carousel costs two dollars.**

Answers

1. False. Not even close. The actual number is more than 25 million annual visitors, according to the Central Park Conservancy.

2. True. Among the most famous are Balto, the heroic sled dog; Alice in Wonderland on a mushroom; and Hans Christian Andersen.

3. False. The duo's 1981 reunion ranks fifth with an estimated crowd of 400,000. Number one was Earth Day 1990, with an estimated crowd of 750,000. That is followed by an antinuclear rally in 1982, a 1991 Paul Simon concert, and a 1993 Luciano Pavarotti recital—all free of course.

4. True. The park issues 2,000 film permits every year, with most going to commercial photographers. Among the most famous Central Park films are *Annie Hall*, *Barefoot in the Park*, *Breakfast at Tiffany's*, *Kramer vs. Kramer*, *Ghostbusters*, and *Wall Street*.

5. False. A ride on the beloved ponies will set you back $1.25 for a three-and-a-half-minute ride, but the second ride is free!

DON'T KNOW MUCH ABOUT *Chicago*

HOME TO THE Potawatomi Indians for some 5,000 years, Chicago was incorporated as a village in 1833 when its population stood at 150 people. (Chicago was incorporated as a city in 1837.) What else do you know about the Second City or the "City of Big Shoulders"? (No Cubs jokes allowed!)

1. Where does the city's name come from?

2. Who were the first whites to reach the Chicago area some 335 years ago?

3. Who started the first permanent settlement in Chicago?

4. What other Chicago "birth" took place 200 years ago?

5. Why is it called the "Windy City"?

6. During World War II, what world-changing event took place in Chicago?

7. What novel about the city's meat industry led to the Pure Food and Drug Act?

Answers

1. The name Chicago comes from the Potawatomi word "Checagou." Peaceful and prosperous hunters and farmers, the Potawatomi welcomed the first whites to arrive.

2. Probably the French-Canadian explorer Louis Jolliet and a Jesuit priest Jacques Marquette. They arrived in 1673 on their way to Canada.

3. Jean Baptiste Point du Sable, a black fur trader from New Orleans who established a trading post in the 1770s. His business became the center of a permanent Chicago settlement.

4. In 1803, a small military post called Fort Dearborn was built.

5. During the 1893 Columbian Exposition, city residents bragged so much that a New York journalist called Chicago the Windy City. The gusts that blow off Lake Michigan made the nickname stick.

6. The first nuclear chain reaction, leading to the development of the atomic bomb, was set off at the University of Chicago on December 2, 1942.

7. Upton Sinclair's muckraking book, *The Jungle* (1906).

DON'T KNOW MUCH ABOUT *West Point*

ON MARCH 16, 1802, Congress established the U.S. Military Academy at West Point, New York, the oldest continuously occupied military post in America. West Point's history dates to the Revolution, when General George Washington considered this spot the most significant strategic position in America. A fort was built on the heights above the Hudson River and a 150-ton chain was stretched across the river to control water traffic. After the Revolution, Washington and other leaders recognized the importance of an institution devoted to the art and science of war and the U.S. Military Academy was born. What else do you know about this American institution?

1. **During the Revolution, West Point's commander attempted to turn the fort over to the British. Who was he?**
2. **Which notable 1829 graduate was later Superintendent of the Academy and then president of Washington & Lee University?**
3. **Which 1950 graduate went "very far" in life?**
4. **What are the chief requirements for admission to West Point?**
5. **What West Point distinction is held by Andrea Lee Hollen?**

ANSWERS

1. Benedict Arnold. In command of West Point in 1780, Arnold worked out a plan to surrender the fort to the British. The plot was foiled and Arnold became a brigadier general in the British Army, and as a British officer led expeditions that burned Richmond, Virginia, and New London, Connecticut.

2. Robert E. Lee, who also served as General in Chief of the Confederate Army during the Civil War.

3. Astronaut Frank Borman, commander of the first flight around the moon. Other West Point astronauts include Buzz Aldrin, Ed White, who died in the Apollo fire in 1967, and Michael Collins, pilot during the first moon landing.

4. A candidate for nomination to the school must be an unmarried U.S. citizen, at least 17 years old and not yet 23 years of age on July 1 of the year of admission; able to meet the school's rigorous academic, physical, and medical requirements; free of legal obligation to support a child or children. Approximately three quarters of the vacancies for the academy are nominated by U.S. senators and representatives, the President, or Vice President.

5. Hollen, Class of 1980, was the first woman graduate of the Academy; she is also a Rhodes Scholar. Women were first admitted to the academy in 1976.

DON'T KNOW MUCH ABOUT *Greece*

Your school books said that Western civilization started about 2,500 years ago in Greece. In ancient times, the Greeks established the arts, philosophy, and science that became foundations of Western culture. Their traditions of justice and individual freedom are basic to democracy. But for most of us, the lesson stopped about 2,000 years ago and modern Greece is a mystery. In Greece, March 25 is celebrated as Independence Day, marking the day in 1821 when the Greeks began a war to throw off the rule of the Ottoman Empire, which was finally won in 1829. Do you think the people of Greece are called "Grecians" and Homer is just Bart Simpson's father? Test your Greek IQ with this quick quiz.

True or False?

1. **Greece is mostly composed of many small islands.**
2. **The capital and largest city in Greece is Athens.**
3. **Although it is on the Mediterranean, Greece is a member of the North Atlantic Treaty Organization (NATO).**
4. **Ithaca, the home of Ulysses, hero of the *Odyssey*, is Greece's largest island.**
5. **Today, Greece is a largely Catholic country.**

ANSWERS

1. False. While, no part of Greece is more than eighty-five miles from the sea, Greece's hundreds of islands together make up about 20 percent of the country. The Greeks have always been seafaring, and many of its legends, including those about Ulysses and Jason, center on sea voyages.

2. True. About 30 percent of all Greeks live in the capital city of Athens or its suburbs. One of the world's most historic cities, Athens attracts more than 90 percent of the tourists who come to Greece (it was the site of the 2004 Summer Olympics). It is famed for the Acropolis, which includes the beautiful ruins of the Parthenon and other ancient temples set on a rocky hilltop. Severe air pollution has damaged the city's ruins and Athens has banned cars from certain sections of the city.

3. True. Following a civil war, Greece joined NATO in 1952 and allowed the United States to set up military bases on its territory in 1953 in an effort to counter Communist guerillas.

4. False. Crete is the largest Greek island. Its major tourist attraction is the famous ruins of Knossos, the center of the ancient Minoan civilization, home of the legendary King Minos.

5. False. About 98 percent of Greece's people belong to the Greek Orthodox Church, the nation's official religion, but everyone has freedom of worship. The Greek government pays the salaries of Greek Orthodox clergy.

DON'T KNOW MUCH ABOUT *Texas*

IT IS BIG—the second largest state after Alaska. According to the 2007 U.S. Census Bureau Estimate, it is populous, with the second largest population after California. And it is proud of its history and tradition as a one-time independent republic. Texas entered the nation on December 29, 1845, as the twenty-eighth state. What else do you know about the "Lone Star State"? Take this quick quiz.

1. **What does "Texas" mean?**
2. **How long was Texas a republic before becoming a state?**
3. **Who was the first President of the Republic, first U.S. Senator from Texas, and later the Governor of the state?**
4. **Which six flags have flown over Texas?**
5. **Which two Presidents were born in Texas?**

ANSWERS

1. The state's name is derived from *tejas*, a Spanish version of an Indian word meaning "friends."

2. Nine years. Independence from Mexico was won in 1836.

3. The hero of the Texas revolution, Sam Houston served two separate terms as President of the Texas Republic, became its first Senator in 1845, and was elected governor of Texas in 1859, but when Texas seceded from the Union, he opposed secession and was deposed by the secession convention.

4. The Spanish, French, Mexican, Republic of Texas, American, and Confederate flags.

5. Dwight D. Eisenhower and Lyndon B. Johnson.

DON'T KNOW MUCH ABOUT *Dodge City*

In Kansas, Dodge City Days celebrates the romance of the Old West each summer with rodeos and country music. But it probably wasn't so appealing in the late nineteenth century. Given legendary status by tabloid journalists and countless Hollywood versions of the cow town legend, the notorious Dodge City of the Old West was a city of bourbon, brothels, and buffalo hides stacked on the town's Front Street, and inspiring the term "stinker." After the buffalo trade ended, the cattle drives began and Dodge really boomed until 1886 when the cattle trade ended and Kansas wheat took over. What else do you know about the "sin city" of the Old West? Take this quick quiz.

1. **Why was Dodge City established five miles from Fort Dodge?**
2. **What was Dodge City's first major product?**
3. **Where does the name "Boot Hill" come from?**
4. **Were famous Hollywood lawmen Bat Masterson, Doc Holliday, Wyatt Earp, and Matt Dillon all real people?**
5. ***Gunsmoke* lasted longer than Dodge City's wild days. How long did this classic television Western run?**
6. **Where did "red light district" originate?**

ANSWERS

1. The commander of the fort, built in 1865 during the Indian wars, prohibited alcohol, so the town was set up to serve the thirsty cowpokes cheap whiskey.

2. Buffalo. More than 850,000 buffalo hides were shipped from Dodge in a three-year period. Famed "lawman" Wyatt Earp had once been a buffalo hunter.

3. Dodge's famous cemetery was for those who couldn't afford to be buried in the more respectable Fort Dodge cemetery. Most were men who "died with their boots on," a phrase that meant dying violently in a brawl or by hanging.

4. Except for *Gunsmoke* Marshal Matt Dillon, all of the others were real people, although much removed from their glorified Hollywood images.

5. *Gunsmoke* ran for twenty seasons. Dodge City's peak lasted from 1872 to 1886.

6. Train masters would carry their red caboose lanterns when they went to visit the town's "soiled doves," or barroom prostitutes.

DON'T KNOW MUCH ABOUT
The Berlin Wall

WINSTON CHURCHILL ONCE said that Communism had dropped an "Iron Curtain" across Eastern Europe. On August 12, 1961, that Iron Curtain became concrete when the Berlin Wall was begun. This Cold War barrier was built to divide the two parts of Berlin—Communist East Berlin and non-Communist West Berlin—and prevent East Germans from escaping Communist rule by crossing into West Berlin. After the Wall was completed, more than 170 people died trying to cross it, most shot by border guards. What else do you know about this fallen symbol of collapsed Communism? Take this quick quiz.

1. **Who controlled the divided city of Berlin after World War II?**
2. **How long was the completed Berlin Wall?**
3. **In a famous speech, who said, "Ich bin ein Berliner" ("I am a Berliner")?**
4. **When was the Berlin Wall destroyed?**

Answers

1. The city was divided into four sectors. The western sector was controlled by the Americans, British, and French; the Soviets controlled the eastern sector.

2. A system of heavily fortified barriers with pipes, barbed wire, and other obstacles installed on top, the Wall was about 26 miles (42 kilometers) long. Walls and other barriers totaling about 110 miles (160 kilometers) were built around the rest of West Berlin.

3. President John F. Kennedy, expressing support for the people of Berlin in a 1963 speech given in West Berlin. Although the sentiment of his speech was understood by everyone, it was later revealed that a "Berliner" is actually the name of a popular breakfast pastry—so Kennedy said the equivalent of "I am a jelly doughnut."

4. Rising demands for freedom burst out in East Germany in 1989. The East Germans opened the Wall in November 1989 and soon began to tear it down. Parts of the Wall were sold to raise hard currency, but several sections remain as memorials. In October 1990, East and West Germany were united into the single non-Communist country of Germany and Berlin was reunited into a single city.

DON'T KNOW MUCH ABOUT *Hawaii*

ALOHA! THE FIFTIETH STATE—the only one that was once a kingdom with its own monarch—entered the Union on August 21, 1959. America's piece of Pacific Paradise, Hawaii does not have a very "pacific" history. In 1893, Queen Liliuokalani, the last constitutional monarch of Hawaii, was overthrown with the aid of U.S. Marines. The Republic of Hawaii was declared with businessman Sanford Dole as President. In 1898, Hawaii was taken over by the United States and became an American territory in 1900. Of course, the Japanese attack on the U.S. base at Pearl Harbor in 1941 drew America into the Second World War. What else do you know about the fiftieth state ? Take this quick quiz.

TRUE OR FALSE?

1. **The first European to reach Hawaii was Magellan, on his famed trip around the globe in 1521.**
2. **The first English name for the Hawaiian Islands was the Sandwich Islands.**
3. **South Point, on the island of Hawaii, is the southernmost point in the United States.**
4. **The Hawaiian Islands are coral atolls.**

ANSWERS

1. False. The British sailor Captain James Cook is generally credited as the first European to reach Hawaii in 1778, although a Spanish captain may have been there earlier. Cook was killed in Hawaii by a native chief in 1779.
2. True. Cook named the islands after his patron and sponsor, John Montagu, the Earl of Sandwich, famed otherwise as the man who stuck meat between two slices of bread.
3. True. South Point lives up to its name.
4. False. All of Hawaii's land is volcanic in origin.

DON'T KNOW MUCH ABOUT *Alaska*

THEY CALLED IT "Seward's Folly" and "Seward's Icebox." Secretary of State William Seward was roundly mocked in 1867 when he brokered a deal to purchase the future forty-ninth state from Russia for $7.2 million, about 2 cents an acre. Buying Alaska was the land deal of the century, right up there with the Louisiana Purchase. What else do you know about America's northern exposure? Take this quick quiz.

TRUE OR FALSE?

1. **Captain James Cook, the first Englishman to reach Hawaii, was also the first Englishman to see Alaska.**
2. **Alaska became the forty-ninth state in 1946.**
3. **The word Alaska means "frozen paradise."**
4. **Alaska's area of more than 591,000 square miles is one-fifth the size of the entire lower forty-eight states.**
5. **Parts of Alaska are only 2.5 miles from Russia.**
6. **The native people of Alaska are called Eskimos.**
7. **North America's tallest mountain is in Alaska.**

ANSWERS

1. True. On his 1776 voyage in search of a Northwest Passage, Cook sailed to Alaska.
2. False. Alaskans voted for statehood in 1946, but it became a state in January 1959.
3. False. Alaska is derived from the Aleut and Inuit words meaning "mainland."
4. True.
5. True. Little Diomede Island is 2.5 miles from Russia's Big Diomede Island.
6. False. "Eskimo" is a French word for the tribe more accurately known as Inuit; there are three other principal tribal groups in Alaska: the Aleut, Tlingit-Haida, and Athabasca.
7. True. Mount McKinley in Denali National Park is 20,320 feet tall.

DON'T KNOW MUCH ABOUT *Australia*

IF YOU ARE like most "Yanks," your knowledge of Australia tends to be limited to koalas, kangaroos, and reruns of *Crocodile Dundee* and *The Road Warrior.* Wondering if Australia is the world's smallest continent or the largest island? Don't know a dingo from a didgeridoo? Take this quick quiz about the "Land Down Under."

TRUE OR FALSE?

1. Sydney, the largest city in Australia, is also the nation's capital.
2. As in New England, the first English settlers in Australia were religious pilgrims seeking freedom of religion.
3. Many of Australia's native Aborigines, like the natives of North America, were killed or died after the arrival of European settlers.
4. A "didgeridoo" is the Australian nickname for a baby kangaroo.
5. The vast interior of Australia is sparsely settled desert or dry grassland.
6. Australia is the world's largest island.

ANSWERS

1. False. With a population around 4.3 million(2006 estimate), Sydney is the largest city, but Canberra is the seat of government.

2. False. Australia served as a penal colony for Great Britain beginning in 1788, after the newly independent America no longer served that purpose.

3. True. When English settlers arrived in 1788, there were about 750,000 indigenous people in Australia. There are now some 250,000 indigenous people, many of them of mixed ancestry.

4. False. A didgeridoo is a traditional aboriginal musical instrument made of bamboo.

5. True. Most Australians live along a thin coastal strip, primarily on the southeastern part of the continent.

6. False. Strictly speaking, Australia is an "island continent." But it is not counted officially among the world's islands, even though it is larger than the largest, which is Greenland, followed by New Guinea, Borneo, Madagascar, and Baffin. Many geographers refer to Australia as part of a larger "continent" called Oceania, which includes some 10,000 islands in the Pacific, among them New Zealand, Indonesia, and the many islands of Polynesia, Micronesia, and Melanesia.

DON'T KNOW MUCH ABOUT *California*

FROM DEATH VALLEY and Silicon Valley and Napa Valley to Valley Girls, California is a special state—of mind. America's most populous state, which was admitted into the Union in 1850 following the Mexican War, discovered gold in 1848 and experienced the Gold Rush in 1849. America's third largest state (after Alaska and Texas) has been attracting dreamers for centuries. What do you know about the "Golden State"? Try this quick quiz.

1. **Where does the word California come from?**
2. **Who were the first Europeans to settle in California?**
3. **The 1930s influx of migrant workers into California inspired what novel by which native Californian?**
4. **If California has one great fault, what is it?**

ANSWERS

1. Spanish discoverer Hernán Cortés first used it in 1535, but no one is sure what it means. Two possibilities: it is drawn from a 1500 Spanish romance in which California is an island near Paradise, or it is derived from the Spanish words for "hot furnace." Sounds like it is either Heaven or Hell.
2. The Spanish Franciscan Father Junipero Serra set up a string of missions, beginning with one at San Diego in 1769. The native inhabitants were Christianized, but were soon used as forced labor at the missions.
3. John Steinbeck's *The Grapes of Wrath*.
4. The San Andreas Fault is the geological fault line running the length of the state and which is responsible for thousands of earthquakes, most of them mild, which strike the state each year.

DON'T KNOW MUCH ABOUT *Canada*

"O CANADA," the opening lyrics of the Canadian anthem, are the sum of what many Americans know about the friendly giant to the North. Celebrated on July 1, Canada Day (once known as Dominion Day) marks the date in 1867 when Canada became a nation, peacefully accomplishing what the American Revolution had done in 1776. What do you know about the world's second largest country (after Russia)? Introduce yourself to America's next-door neighbor with this quick quiz.

TRUE OR FALSE?

1. **The word "Canada" means "large and cold."**
2. **Canada's oldest city is Quebec.**
3. **The capital of Canada is Toronto.**
4. **Canada is now comprised of ten provinces, two territories, and one native homeland called Nunavut.**
5. **America has never fought with Canada.**

Answers

1. False. Canada is an Algonquin word for "village."
2. True. It was established by the French in 1608. "New France," France's American empire, was surrendered to the British in 1763. About one quarter of the country's population is French-Canadian.
3. False. Ottawa is seat of the government; Toronto is the largest city.
4. True. The provinces are Alberta, British Columbia, Manitoba, New Brunswick, Newfoundland, Nova Scotia, Ontario, Prince Edward Island, Quebec, and Saskatchewan. The two territories are the Northwest and Yukon. Nunavut ("Our Land") was established as an Inuit homeland in April 1999.
5. False. During the War of 1812, while Canada was still a colony of Great Britain, American troops invaded. York (Toronto) was burned and several naval battles were fought on the Great Lakes.

DON'T KNOW MUCH ABOUT *Presidential Libraries*

AS OF 2008, there were seventeen presidential libraries, with the openings of President Clinton's in Little Rock, Arkansas, in November 2004, and the Lincoln Museum and Library in Springfield, Illinois, between 2004 and 2005. Construction of the next presidential library, that of George W. Bush, is planned to begin in 2009 in Dallas, Texas, at Southern Methodist University. The first presidential library, that of Franklin D. Roosevelt in Hyde Park, New York, was opened in November 1939. Presidential libraries collect documents and other items associated with a former president of the United States, and these libraries function as resources for researchers. But most also have museums open to the public. What do you know about these unique American institutions?

1. **Which of the presidential libraries are administered by the federal government's National Archives and Records Administration?**
2. **Which presidential libraries are not administered by the National Archives?**
3. **Which President's unusual records are also maintained by the National Archives?**
4. **True or False? Presidential libraries are located in the president's birthplace.**
5. **Who pays for presidential libraries?**

ANSWERS

1. Twelve of them are administered by NARA. In presidential order with locations, they are Herbert Hoover (West Branch, Iowa); Franklin D. Roosevelt (Hyde Park, New York); Harry S. Truman (Independence, Missouri); Dwight D. Eisenhower (Abilene, Kansas); John Fitzgerald Kennedy (Boston); Lyndon Baines Johnson (Austin, Texas); Richard Nixon (Yorba Linda, California); Gerald R. Ford (Ann Arbor, Michigan); Jimmy Carter (Atlanta, Georgia); Ronald Reagan (Simi Valley, California); George Bush (College Station, Texas); and William J. Clinton (Little Rock, Arkansas).

2. The Rutherford B. Hayes Library in Freemont, Ohio, is owned by the state of Ohio and operated by the Rutherford B. Hayes Presidential Center. The Abraham Lincoln Presidential Library and Museum is operated by the state of Illinois. The newest of the presidential libraries, the Lincoln Presidential Library opened to the public on October 14, 2004. The Presidential Museum opened over four days in April 2005. The William McKinley Museum in Canton, Ohio, is operated by a private nonprofit group. The Calvin Coolidge Museum is in the Forbes Library in Northampton, Massachusetts. The Woodrow Wilson Library is operated by a nonprofit foundation in Staunton, Virginia.

3. The official records of the presidency of Richard M. Nixon. These records, which include the Oval Office tapes from the Watergate controversy, are kept in Alexandria, Virginia.

4. False. The Wilson, Hoover, Roosevelt, and Nixon libraries are each in their place of birth.

5. The libraries are built with private funds but then operated with federal funds by the National Archives.

HISTORIC HAPPENINGS AND CIVICS

DON'T KNOW MUCH ABOUT *Three Mile Island*

IN THE EARLY HOURS of March 28, 1979, a reactor at the Three Mile Island nuclear power facility overheated, releasing radioactive gases. As thousands of residents fled to emergency shelters during a tension-packed week, plant workers frantically tried to prevent the nightmare of a complete nuclear meltdown. No injuries due to radiation occurred at Three Mile Island, but the equipment failure, human error, and bad luck had combined to create America's worst nuclear accident. What do you know about the incident that changed American energy policy?

1. **Where is Three Mile Island?**
2. **What is *The China Syndrome*?**
3. **Which nation has the most nuclear power plants?**
4. **Which country gets most of its electricity from nuclear power?**
5. **What was the most deadly nuclear power incident in history?**

Answers

1. It is an island in Pennsylvania's Susquehanna River near Harrisburg. Nuclear reactors are usually located near water because they need large quantities of water for cooling purposes.

2. The Three Mile incident occurred three days after the movie *The China Syndrome* opened. The title refers to the concept that if an American nuclear plant melts down, it will melt through the Earth until it reaches China.

3. The United States, with 103 operating plants in 2007.

4. According to the International Atomic Energy Agency (2007), France is the leader with 78 percent; the United States gets about 19 percent of its electricity from nuclear plants. Nuclear power reactors produce electric power in more than thirty countries and there are more than 435 reactors worldwide. They produce about 15 percent of the world's electric power. In 2007, twenty more plants were under construction.

5. The worst nuclear accident occurred in 1986 at the Chernobyl nuclear power plant near Kiev in Ukraine (then part of the Soviet Union). Soviet officials reported that 31 people died from radiation sickness or burns and more than 200 others were seriously injured. Experts believe that the radiation released could lead to tens of thousands of cancers in nearby countries.

DON'T KNOW MUCH ABOUT *A Century of Flight*

AFTER THE WRIGHT Brothers made their pioneering flight in 1903 (see page 41), air travel took some time to catch on. But the Wright Brothers had changed the world forever. There have been many other milestones during more than a century of flight. Straighten your set back and put up your tray table for this quick quiz.

1. **What body of water did Frenchman Louis Blériot cross in a famed 1909 flight?**
2. **What did the Post Office issue in 1918 for the first time?**
3. **In 1929, what aviation first was performed by registered nurse and student pilot Ellen Church?**
4. **What regular air service was initiated in 1958 by both the now defunct Pan Am and BOAC?**
5. **What large service did Pan Am introduce in 1970?**

ANSWERS

1. He crossed the English Channel from Calais in 37 minutes.
2. Airmail stamps, as the U.S. government used Army pilots to begin the first permanent airmail service. The first official stamp showed a biplane in flight, but was printed upside down, making the stamp a rare collectors' item.
3. She was the first flight attendant, working for newly formed United Airlines. She was told to hire seven other nurses—aged 25, single, not more than five feet four or heavier than 115 pounds. The attendants are credited with allaying public fears of flying.
4. Transatlantic jet service. Within a year, more than 1.5 million passengers were "crossing the pond," nearly double the number who went by sea.
5. Pan Am introduced "jumbo jets" with the Boeing 747.

DON'T KNOW MUCH ABOUT *The Grenada Invasion*

America has had a very mixed success with "regime change" in the past. But one of the most successful of these undertakings happened in 1983 when the United States invaded tiny Grenada. One of the smallest independent countries in the world, Grenada had been seized in a Marxist military coup on October 19, 1983. When Cuban forces began building a military airstrip on the island, President Reagan decided to act. In the early morning of October 25, some 1,200 troops assaulted the island in Operation Urgent Fury. The invasion force grew to more than 7,000 and the island's defenders soon surrendered or fled into the mountains. What else do you know about this island nation and its "liberation?"

True or False?

1. **Grenada first became independent in 1784, a few years after America.**

2. **The stated reason for the American invasion was the safety of Americans on the island.**

3. **Grenada marked the first time reporters were "embedded" with the military.**

4. **The Grenada invasion came within days of a major American military disaster.**

5. **There were no American combat deaths in the Grenada invasion.**

ANSWERS

1. False. Grenada became independent of Great Britain in 1974, which had held the island for nearly 200 years.

2. True. The presence of nearly 1,000 American medical students in Grenada was cited as the reason for the American action. But the bigger reason was the desire to eliminate a Marxist government and deal a setback to Fidel Castro's Cuba. After U.S. forces captured the coup leaders and hundreds of Cuban advisers, elections were held the following year.

3. False. The Grenada invasion was carried out with a complete press blackout. No reporters arrived for several days.

4. True. On October 23, 1983, a suicide bomber destroyed the Marine barracks in Beirut, Lebanon, killing 240 U.S. Marines.

5. False. Officially, nineteen Americans died along with forty Grenadans and twenty-nine Cubans.

DON'T KNOW MUCH ABOUT *The Battle of Saratoga*

THINK SARATOGA. If you do, New York, mineral water, and horse racing may come to mind. But if not for events at Saratoga in 1777, we might still be singing "God Save the Queen." Saratoga was the scene of several battles that changed the course of American independence and history. Fought in September and October 1777, the fighting stopped a British army marching from Canada to crush the Revolution. This American victory also brought France into the war as America's ally, and without French guns, ships, and men, the Revolution might have been lost. What do you know about this and several other turning points that changed the course of the Revolution?

TRUE OR FALSE?

1. **There's a statue of turncoat Benedict Arnold's boot at the Saratoga Battlefield.**
2. **The "shot heard 'round the world" was fired at Bunker Hill in Boston.**
3. **The Battle of Bunker Hill (June 17, 1775) was an American loss and wasn't even fought at Bunker Hill.**
4. **After a string of losses, the Americans won the key battle of Charleston, South Carolina, in May 1780.**
5. **The decisive American victory at Yorktown was largely made possible with French help.**

ANSWERS

1. True. Arnold, who later betrayed the American cause, was a hero at Saratoga. He was wounded in the leg, and the statue commemorates his role in the victory there.

2. False. The famed shot that began the war was fired in April 1775 at Lexington, Massachusetts.

3. Both true. While the American troops officially lost the battle, it was a costly victory for the British who sustained high casualties in taking Breed's Hill, not Bunker Hill.

4. False. The battle there was the single worst American loss of the Revolution.

5. True. A combined French and American force of about 18,000 soldiers and sailors surrounded British General Cornwallis at Yorktown. About 5,500 French soldiers had reached America in July 1780 led by Lieutenant General Jean Rochambeau. They were aided by a large French fleet under Admiral Francois de Grasse.

DON'T KNOW MUCH ABOUT
The Russian Revolution

On November 7, 1917, history changed radically when workers, soldiers, and sailors led by the Bolsheviks took over Russia's government. Known as the October Revolution because the date was October 25 under the old Russian calendar, this was the second of two uprisings in Russia that year. Earlier, in March, the people had deposed the ruling czar. In 1922, the Communist government created a new nation called the Union of Soviet Socialist Republics (U.S.S.R.). Who's who in the Russian Revolution? Take this quick quiz.

1. **What German philosopher's ideas form the basis for modern Communism?**

2. **Who founded the Communist Party in Russia, led the October Revolution of 1917, and then ruled the country until his death in 1924?**

3. **What became of the last Russian czar?**

4. **Who was the Siberian healer and holy man who served as an adviser to the czar and contributed to his downfall?**

5. **Which Soviet leader played a minor role in the revolution, then had history books rewritten to say that he led the revolution?**

6. **Who is the American journalist best known for his book *Ten Days That Shook the World* (1919), an eyewitness account of the Russian Revolution?**

Answers

1. Karl Marx (1818–1883). His pamphlet *The Communist Manifesto* appeared in 1848.

2. Vladimir Lenin (1870–1924).

3. Nicholas II gave up the throne on March 15. He and his family were imprisoned and almost certainly shot to death in July 1918.

4. Rasputin, the "mad monk." A group of the czar's supporters poisoned and shot Rasputin, then threw him into the Neva River, where he drowned.

5. Joseph Stalin (1879–1953).

6. John Reed, subject of the film *Reds* and founder of the American Communist Party. Reed died of typhus in Russia and was buried in front of the Kremlin.

DON'T KNOW MUCH ABOUT *The Cuban Missile Crisis*

IN OCTOBER 1962, America and the Soviet Union came frighteningly close to all-out nuclear war. A dispute over Soviet missiles based in Cuba, only ninety miles from U.S. soil, led to a confrontation that reached its peak during thirteen days that had President Kennedy and other American leaders envisioning the worst. (It also inspired the film *Thirteen Days,* which takes some liberty with the facts, in typical Hollywood fashion.) What do you know about this dangerous moment in Cold War history?

TRUE OR FALSE?

1. **The crisis began when Russian cargo ships started to unload missiles disguised as wheat in Havana.**
2. **To prevent the missiles from being deployed, President Kennedy ordered a naval blockade, or "quarantine," of Cuba.**
3. **The missiles in Cuba were a Soviet response to U.S. missile bases in Turkey.**
4. **The disastrous Bay of Pigs invasion took place one year after the Cuban Missile Crisis.**

ANSWERS

1. False. American intelligence photographs showed the construction of Soviet nuclear missile installations on the island. These photos were then displayed at the United Nations.

2. True. The blockade was of all military equipment being shipped to Cuba, and the Soviet ships turned back before any actual confrontation.

3. True. In 1959, American missiles had been deployed near the Soviet border and were considered a great threat by the Soviet Union, which feared an American "first strike." These missiles were removed as part of the negotiations between Kennedy and the Soviet leader Khruschev.

4. False. The CIA-led Bay of Pigs invasion to overthrow Fidel Castro took place in April 1961, more than a year before the Missile Crisis. In 1962, the United States was still actively planning another invasion of Cuba.

DON'T KNOW MUCH ABOUT
The Battle of Bataan

IN 1942, IN THE DARKEST early days of World War II, American troops suffered a loss that remains one of the most ghastly episodes in American military history. After Japan attacked Pearl Harbor, United States and Allied Filipino troops were holding out in the Philippines against a large Japanese army. Cut off from help, this band of defenders beat back Japanese attacks for more than three months. But on April 9, 1942, when Japanese forces broke through, the American commander surrendered. Some of the soldiers escaped to the fortress of Corregidor and held out until May 6, 1942. The prisoners were then forced into what became infamously known as the Death March. In 1954, the battlefield areas of Bataan and Corregidor were made into Philippine national shrines. What do you know about this wartime tragedy?

1. **Where is Bataan?**
2. **Who was the American commander in the Philippines?**
3. **What happened on the Bataan Death March?**
4. **When did the United States retake the Philippines?**

Answers

1. Bataan Peninsula juts into Manila Bay from the southwestern coast of Luzon, the largest of the Philippine Islands.

2. General Douglas MacArthur. President Roosevelt ordered MacArthur to go to Australia, and he left the Philippines in March 1942, promising the Filipinos, "I shall return." Lieutenant General Jonathan M. Wainwright took command of United States and Filipino forces in the Philippines.

3. The Japanese took about 75,000 Americans and Filipinos prisoner. Most of them were forced to march nearly 100 miles through jungle to prisoner of war camps. During this dreadful forced march, thousands of American and Filipino prisoners died of disease and mistreatment. Many more then died in captivity.

4. In February 1945, troops under the command of MacArthur returned to Bataan. The Japanese forces surrendered, and the American and Filipino survivors, including an emaciated General Wainwright, were freed.

DON'T KNOW MUCH ABOUT *The Battle of Yorktown*

"THESE ARE THE TIMES that try men's souls," wrote Thomas Paine in 1776, when things looked rather bleak for the American Revolutionary cause. Five years later, on October 19, 1781, the last major battle of the Revolutionary War was fought at Yorktown, Virginia, on Chesapeake Bay. George Washington's army delivered a crushing defeat to British forces. What do you know about this turning point in American history?

TRUE OR FALSE?

1. **One of the British commanders in Virginia in 1781 was traitor Benedict Arnold.**
2. **The American victory at Yorktown was largely made possible with French help.**
3. **The Americans defeated a British force of more than 50,000 at Yorktown.**
4. **Britain's loss at Yorktown did not end the Revolutionary War.**
5. **The war was officially ended by the the Treaty of Versailles.**

ANSWERS

1. True. In January 1781, Arnold led raids in Virginia for the British, who had made him a brigadier general.

2. True. A combined French and American force of about 18,000 soldiers and sailors surrounded British General Cornwallis at Yorktown. About 5,500 French soldiers had reached America in July 1780 led by Lieutenant General Jean Rochambeau. They were aided by a large French fleet under Admiral Francois de Grasse.

3. False. More than 8,000 men, about one quarter of Britain's military force in America, surrendered as a British band reportedly played a tune called "The World Turned Upside Down."

4. True. In some areas, the fighting dragged on for two more years, but the defeat at Yorktown brought a new group of British ministers to power early in 1782. They began peace talks with the Americans.

5. False. It was the Treaty of Paris, negotiated in 1782. Benjamin Franklin, John Adams, and John Jay represented the United States.

DON'T KNOW MUCH ABOUT *The Gulf War*

PRESIDENT BUSH. Dick Cheney. Colin Powell. Although it is a different President Bush, these names from headlines in the twenty-first century recall events of the early 1990s, when the U.S. led an Allied effort against Iraq in the Persian Gulf War. After Iraq invaded neighboring Kuwait on August 2, 1990, the United Nations ordered Saddam Hussein to withdraw his troops. When a January 15, 1991, deadline was ignored by Saddam Hussein, a massive air attack began on January 16, 1991. The U.S. led a coalition of thirty-two nations in this military effort. What else do you recall about the first war against Saddam Hussein's Iraq?

1. **In what positions did Dick Cheney and Colin Powell serve during the Persian Gulf War?**
2. **What is a "scud"?**
3. **How long did the ground offensive to retake Kuwait last?**
4. **What was the ancient name for the modern land of Iraq?**

ANSWERS

1. Vice President Cheney served as U.S. Defense Secretary; Secretary of State Colin Powell was Chairman of the Joint Chiefs of Staff.

2. A missile developed by the Soviet Union and used by Iraq during the war. Although crude in technological terms, the Scuds caused great fear when they were fired at Israel and Saudi Arabia and were thought to possibly carry biological or germ weapons.

3. The assault on Iraq's ground forces began on February 24 and lasted until February 28 when President George H. W. Bush declared a cease-fire.

4. Mesopotamia, which comes from the Greek for "between two rivers," a reference to the Tigris and Euphrates rivers. Often called the "Cradle of Civilization," it was the site of some of the earliest human culture in the world.

DON'T KNOW MUCH ABOUT *Presidential Assassinations*

In presidential history, September has a curious distinction. Two of the four presidents killed by an assassin died that month, although one of them had been shot several months earlier. What do you know about the sad history of presidential assassins?

1. **A Civil War general, which President was shot on July 2, 1881, but died in September 19, 1881? Who was his assassin?**
2. **Another President with a distinguished Civil War record, which President was shot on September 6, 1901, and died on September 14? Who was his assassin?**
3. **What item from the ill-fated box at Ford's Theatre where Lincoln was shot made news in 2001?**
4. **Name the Vice Presidents who have succeeded to the presidency following an assassination.**

ANSWERS

1. Twentieth President James A Garfield, shot in a train station. His assassin, Charles Guiteau, was a deranged office seeker who wanted a government job. Garfield lingered painfully for more than two months before he succumbed to blood poisoning.

2. Twenty-fifth President William McKinley, who was shot twice at close range while shaking hands at an Exposition in Buffalo, New York. He seemed to recover from his wounds but then died. Unemployed millworker and anarchist Leon Czolgosz shot McKinley.

3. A flag that was draped in Lincoln's box was discovered in storage at the Connecticut Historical Society.

4. Andrew Johnson (Lincoln), Chester A. Arthur (Garfield), Theodore Roosevelt (McKinley), and Lyndon B. Johnson (Kennedy).

DON'T KNOW MUCH ABOUT *Woodstock*

ISN'T IT IRONIC? The original Woodstock festival happened in 1969 in a cornfield. The version that took place in the summer of 1999 was on the abandoned Griffis Air Force base. For three days in August 1969, a crowd estimated at between 300,000 and 500,000 people braved the elements to become part of rock and roll history and modern American lore. The later edition of Woodstock offered cellular phone access and cash machines. What do you know about the first "Three Days of Peace and Music"?

TRUE OR FALSE?

1. The first Woodstock festival took place in the hamlet of Woodstock, in upstate New York.

2. The event was held on the property of one Max Yasgur, who told a cheering crowd, "I'm a farmer."

3. The American anthem was played by guitarist Jimi Hendrix.

4. As the Rolling Stones played, a fight broke out and a man was stabbed to death.

5. The site of the original festival was paved over for a parking lot.

Answers

1. False. Originally planned for Woodstock, the festival was moved 60 miles away to Bethel, New York, when Woodstock town fathers objected to the planned music festival.

2. True. The concert stage was set up in Yasgur's 400-acre field.

3. True. Hendrix played "The Star-Spangled Banner."

4. False. The Rolling Stones did not play at Woodstock; their ill-fated concert took place in Altamont, California, the following year.

5. False. In July 2006, the Bethel Woods Center for the Arts, a 4,800-seat covered pavilion, opened on the original festival site with a performance by the New York Philharmonic.

DON'T KNOW MUCH ABOUT *The Great Chicago Fire*

EACH YEAR SINCE 1922, Fire Prevention Week has commemorated one of the great disasters in American history. A raging fire that began on October 8, 1871, nearly leveled the entire city of Chicago. In the wake of a severe drought, the fire spread rapidly through old wooden homes in the heart of the city, burning more than 2,100 acres in 27 hours, killing some 300 people, destroying some 17,500 buildings, and leaving 100,000 residents homeless. With aid from around the world, the Windy City was rebuilt in remarkably short time. What else do you know about this great American disaster?

1. **According to local history and legend, what started the fire?**
2. **Which foreign monarch contributed books from a private collection to rebuild the Chicago library's collection?**
3. **What great American document, written by a prominent resident of Illinois and kept in the Chicago Historical Society building, was destroyed in the fire?**
4. **What famous popular song did Chicago's disaster inspire?**

Answers

1. A milk cow belonging to Mrs. Catherine O'Leary supposedly kicked over a kerosene lantern and set the blaze. That story has been dismissed as myth and the actual source of the blaze is a mystery, although it may have started in the barn of a man named Patrick O'Leary.
2. Queen Victoria donated books to rebuild the library.
3. Abraham Lincoln's draft of the Emancipation Proclamation.
4. "There'll be a hot time in the old town tonight."

DON'T KNOW MUCH ABOUT *The Census*

THERE ARE THREE kinds of lies, Mark Twain once said: "Lies, damn lies, and statistics." With that in mind, consider the American census, first taken on August 1, 1790. When Thomas Jefferson delivered the first census results to President Washington, he gave him two sets of figures: the actual tally and Jefferson's estimates of the correct numbers. After more than two hundred years, the accuracy of this national nose count is still a political hot potato. The best way to count all those noses and which ones got missed has been debated ever since. What do you know about the census? Be counted in this quick quiz.

TRUE OR FALSE?

1. **The Constitution requires the census to be taken at the request of Congress to determine tax levels.**
2. **The U.S. population in 1790 was approximately 4 million, including nearly 700,000 slaves.**
3. **Since the 1790 count, New York has always been America's largest city.**
4. **In 2007 the Census Bureau estimated America's population to be more than 305 million.**

ANSWERS

1. False. The constitutional purpose of the census is to determine each state's representation in Congress or apportionment.

2. True. The African population in 1790, including free African Americans, was nearly 20 percent of the population. In 2006 black residents were approximately 13 percent, according to the Census Bureau.

3. False. In the first census, Philadelphia was Number One. New York took over by 1800, and has yet to relinquish the title.

4. True. In October 2007, the Census Bureau estimated the population at 303,241,574. There is one birth every 7 seconds and one death every 13 seconds.

DON'T KNOW MUCH ABOUT
The Twenty-first Amendment

"THE TWENTY-FIRST AMENDMENT" is a popular name for bars around the country. The reason is simple: "The eighteenth article of amendment to the Constitution of the United States is hereby repealed." With those simple words, the Twenty-first Amendment was ratified by the states on December 5, 1933, and ended an era in America. The Twenty-first Amendment repealed the Eighteenth Amendment, ending the policy called Prohibition. It was now legal to drink alcohol in most of America (except for a few localities that remained "dry"). What do you know about Prohibition and its repeal? Toss back this quick quiz.

TRUE OR FALSE?

1. The Eighteenth Amendment to the Constitution was ratified in 1917.

2. During Prohibition, doctors could prescribe alcohol.

3. The chief reason for Prohibition was ending crime.

4. Franklin D. Roosevelt favored the repeal of Prohibition.

5. Some states continue Prohibition.

ANSWERS

1. False. In December 1917, Congress approved the 18th Amendment, but it was ratified by the states in January 1919 and went into effect in 1920. It is the only amendment that has ever been repealed.

2. True. Many physicians gave their patients prescriptions for legal "medicinal" wine or liquor. Wine for religious use, such as communion, was also permitted.

3. False. Many social reformers blamed alcohol for poverty, health problems, and the neglect of wives and children. Employers also felt that drunkenness reduced productivity. Prohibition is actually believed to have created a much larger crime problem because bootlegging and illegal liquor sales became so profitable for such mobsters as Al Capone.

4. True. In the 1932 presidential campaign, the Democratic Party endorsed the repeal of Prohibition, and Roosevelt won by a large margin.

5. False. While a few states kept Prohibition until the 1960s, Mississippi became the last state to repeal statewide Prohibition in 1966.

HOLIDAYS AND TRADITIONS

DON'T KNOW MUCH ABOUT *Tax Day*

"WHAT IS THE DIFFERENCE between a taxidermist and a tax collector?" Mark Twain once asked. "The taxidermist takes only your skin." When that special day in April rolls around, our history proves that balking at taxes is more American than apple pie. Since colonial times, when the Founders protested taxes on stamps, sugar, and tea, to the early days of the Republic when farmers rebelled over a tax on whiskey, the "T" word has irked Americans. Humorist Will Rogers had it right when he said, "The income tax has made more liars out of the American people than golf has." Tax your knowledge of the IRS with this quick quiz.

TRUE OR FALSE?

1. **The first federal income tax in U.S. history was passed by the Continental Congress during the Revolution.**

2. **The current income tax was established in 1919 with passage of the Eighteenth Amendment.**

3. **Before the income tax, government revenue largely came from customs duties.**

4. **The IRS received more than 135 million individual tax returns in 2007.**

Answers

1. False. The first income tax was passed in 1861 to raise funds for the Union war effort during the Civil War. Although killed by Congress in 1872, the income tax was declared unconstitutional in 1881.

2. False. The 16th Amendment, ratified in 1913, authorizes the income tax. The Supreme Court upheld its constitutionality in 1916. The Eighteenth Amendment authorized Prohibition.

3. True. Until 1917, customs duties provided most of the U.S. government's finances. In 1917, tax revenues surpassed customs duties.

4. True. Some 138,894,000 individual returns were filed that year.

DON'T KNOW MUCH ABOUT
Labor Day

NEXT TIME LABOR DAY rolls around, take a moment to thank Peter J. McGuire. The president of the United Brotherhood of Carpenters and Joiners, McGuire was the driving force behind a national holiday in honor of the American worker. Begun in 1882 by the Knights of Labor, a pioneering union, Labor Day was signed into law in 1894 and set on the first Monday in September. What else do you know about labor in America?

1. **In 1938, the Fair Labor Standards Act set the first federal minimum wage. What was it?**
2. **In 1960, about a third of American workers were union members. What was the percentage of union workers in 2005?**
3. **What union was thrown out of the AFL-CIO in 1957 for corruption?**
4. **What distinction does Frances Perkins hold?**

ANSWERS

1. Twenty-five cents per hour for workers engaged in interstate commerce. As of July 24, 2007, it was $5.85, the first increase in a decade. In July 2008, it reached $6.55 per hour and will increase to $7.25 per hour after July 2009.
2. In 2006, 12 percent of wage and salary workers were union members, down slightly from 2005, according to the U.S. Department of Labor. Union membership declined from 20.1 percent in 1983, the first year for which comparable data are available.
3. The Teamsters.
4. Secretary of Labor under Franklin D. Roosevelt and then Harry Truman, Perkins was the first woman cabinet member, appointed in March 1933.

DON'T KNOW MUCH ABOUT *Leap Day*

HOW LONG DOES IT TAKE the Earth to complete an orbit around the Sun? Most people would answer 365 days. That's close, but no cigar. It actually takes a bit longer to make a full trip around the Sun. (By the way, a day is one revolution of the Earth on its axis.) To be precise, it takes 365.2422 days. That's an extra 5 hours, 48 minutes, and 46 seconds. For centuries, those extra hours, minutes, and seconds have caused problems for people who needed accurate calendars to plant crops or celebrate important holidays. Long ago, the western world opted for a simple solution: add one day to the calendar every four years. Now get out your calculators. This extra day now falls on February 29 every four years, except in years which end in 00—unless those years are divisible by 400. That's why 2000 ends in double zero, but it was still a Leap Year. Got that?

TRUE OR FALSE?

1. The Leap Year was created under Julius Caesar in 46 BC.

2. By 1500, the old calendar was no longer accurate, and Pope Gregory XIII ordered a correction to correctly place holidays such as Easter.

3. The "Gregorian Calendar" immediately became the standard accepted by the western world.

4. February 29 was once known as "Sadie Hawkins Day."

ANSWERS

1. True. The Julian Calendar was the western standard for centuries.

2. True. The Julian Calendar had been miscalculated so that holidays such as Easter were "wandering." Under Gregory, the Leap Year was established in its present form and is called the "Gregorian Calendar."

3. False. Most Protestant nations, including England and its colonies, balked at a Roman Catholic calendar. The British and America did not accept the Gregorian Calendar until 1752.

4. False. It used to be called "Bachelor's Day," the one day on which women could acceptably propose to men.

DON'T KNOW MUCH ABOUT *Father's Day*

ALTHOUGH MOTHER'S DAY gets the most attention and sells the most flowers, perfume, and candy, Dads got their due on the third Sunday in June. In honor of Dad, try this Father's Day quiz.

1. **Mother's Day was first celebrated in 1908. When was Father's Day first celebrated?**
2. **What river name comes from the ancient Indian word for "Father of the Rivers"?**
3. **Most American Presidents were fathers. How many American Presidents were childless? And who had the most children?**
4. **Which Presidents were father and son? Which were grandfather and grandson?**
5. **Who is known as the "Father of Angling"?**
6. **The Welsh poet Dylan Thomas wrote his most famous poem for his dying father. What is the title?**

Answers

1. The first recorded Father's Day was celebrated in 1910 in Spokane, Washington.

2. Mississippi.

3. Washington, James Madison, Andrew Jackson, James Polk, James Buchanan, and Warren G. Harding had no legitimate children. John Tyler had fourteen children live to maturity.

4. John Adams and John Quincy Adams were father and son. So are George H. W. Bush and George W. Bush. William Henry Harrison was Benjamin Harrison's grandfather.

5. Famed British author Izaak Walton (1593–1683), author of *The Compleat Angler, or the Contemplative Man's Recreation.*

6. "Do Not Go Gentle Into That Good Night."

DON'T KNOW MUCH ABOUT *Bastille Day*

On July 14, 1789, an angry crowd stormed a state prison in Paris that stood as a symbol of royal tyranny. They surrounded the Bastille in order to seize the gunpowder stored inside. Troops fired on the rebels, but the people overpowered them. The bloody French Revolution had begun. The people of France have come to mark July 14 as their national holiday, the French version of the Fourth of July. What else do you know about this celebration of "Liberty, Equality, and Fraternity"? Try this quick quiz.

True or False?

1. **Bastille means "prison" in French.**
2. **France was declared a republic in 1792 and the king and queen were later executed.**
3. **When told that people were starving, the French queen Marie Antoinette said, "Let them eat cake."**
4. **Thousands of people were executed during the French Revolution's "Reign of Terror."**
5. **The French General Charles de Gaulle eventually took control of the Revolution.**
6. **The Bastille remains one of the most popular tourist attractions in Paris.**

ANSWERS

1. False. Bastille is a French word for a "strongly fortified structure." It was built as a fortress by King Charles V around 1370. By 1789, the Bastille held only a handful of prisoners.

2. True. Louis XVI was beheaded on the guillotine in January 1793; his wife, Marie Antoinette, was guillotined in October 1793.

3. Probably false. The saying was ascribed to the queen as evidence of her disregard for the people's welfare, but it had been recorded much earlier by the French writer Rousseau.

4. True. During the political infighting of the Revolution as many as 18,000 people were guillotined in Paris and many thousands more died elsewhere in France.

5. False. The French Revolution ended in 1799, when the brilliant general Napoleon Bonaparte seized control of the government. Charles de Gaulle was the leader of the Free French who fought the Nazis during World War II.

6. False. The Bastille was torn down.

DON'T KNOW MUCH ABOUT *Flag Day*

In 1877, Congress designated June 14 as Flag Day to commemorate the one hundredth anniversary of the design of the American flag. With its thirteen red and white stripes in honor of the original states, the U.S. flag has changed a lot since then, with fifty stars now representing the states. But the flag has a surprisingly obscure history. How much do you know about the "stars and stripes"?

True or False?

1. **The original design, with thirteen stars in a circle, was the handiwork of seamstress Betsy Ross.**
2. **The American flag is never lowered to honor visiting heads of state.**
3. **The Pledge of Allegiance to the flag, composed in 1776, always included the words "one nation under God."**
4. **It is legal to burn the flag as a form of protest.**

ANSWERS

1. False, probably. The Betsy Ross legend has been discredited. The likely father of the flag was Francis Hopkinson, a member of the Continental Navy Board.

2. True. In a longstanding tradition, the flag is never dipped to any other nation's, including during the Olympics.

3. Double False. The Pledge was composed in 1892 and the words "under God" were added in 1953.

4. True. The Supreme Court has ruled that burning the flag in protest is speech protected under the Fifth Amendment.

DON'T KNOW MUCH ABOUT *The Pilgrims*

IF APRIL SHOWERS bring May flowers, what do Mayflowers bring? Pilgrims! With Thanksgiving around the corner, cutouts of Pilgrims in black clothes and clunky shoes are sprouting all over the place. You may know that the Pilgrims sailed aboard the *Mayflower* and arrived in Plymouth, Massachusetts, in 1620. But did you know their first Thanksgiving celebration lasted three whole days? What else do you know about these early settlers of America? Don't be a turkey. Try this quiz.

TRUE OR FALSE?

1. **Pilgrims always wore stiff black clothes and shoes with silver buckles.**
2. **The Pilgrims came to America in search of religious freedom.**
3. **Everyone on the *Mayflower* was a Pilgrim.**
4. **The Pilgrims were saved from starvation by a native American friend named Squanto.**
5. **The Pilgrims celebrated the first Thanksgiving in America.**

ANSWERS

1. False. Pilgrims wore blue, green, purple, and brownish clothing for everyday. Those who had good black clothes saved them for the Sabbath. No Pilgrims had buckles—artists made that up later!

2. True. The Pilgrims were a group that had broken away from the Church of England.

3. False. Only about half of the 102 people on the *Mayflower* were Pilgrims. The others, called "Strangers," just wanted to come to the New World.

4. True. Squanto, or Tisquantum, helped teach the Pilgrims to hunt, farm, and fish.

5. False. The Indians had been having similar harvest feasts for years. So did the English settlers in Virginia and Spanish settlers in the southwest before the Pilgrims even got to America.

DON'T KNOW MUCH ABOUT
Human Rights Day

INTERNATIONAL HUMAN RIGHTS DAY was established to commemorate the adoption of the United Nations's Universal Declaration of Human Rights on December 10, 1948. Intended as a standard of basic rights for people everywhere, the Declaration lists thirty specific rights—political, social, economic, and cultural—to which people everywhere are entitled. Among them are freedom from slavery and protection from cruel and unusual punishment and torture. What do you know about this international landmark and other notable bills of rights?

1. **What famous American woman helped draft the Universal Declaration?**
2. **Which rock and roll band projected the Declaration of Human Rights during a world tour?**
3. **What is usually considered the oldest "bill of rights"?**
4. **Which of the thirteen colonies crafted the first American bill of rights?**

ANSWERS

1. Former First Lady Eleanor Roosevelt, who chaired the committee that drafted the document.

2. U2 displayed the Declaration, following performances of their song *Running to Stand Still*.

3. The Magna Carta, or Great Charter, approved in June 1215 near London. In it, the king granted rights to the English aristocracy. While ordinary people gained little from it, the Magna Carta became a model for democratic government and individual rights.

4. Virginia, which passed the Virginia Declaration of Rights in June 1776. It is considered the first declaration of human rights in modern times and influenced Thomas Jefferson when he wrote the Declaration of Independence a few months later.

DON'T KNOW MUCH ABOUT
World Health Day

FOUNDED IN 1948 as a specialized agency of the United Nations, the World Health Organization (WHO) follows a mission to build health systems throughout the world, especially in developing countries. WHO helps name and classify diseases, but disease prevention is its key goal. The agency works to provide safe drinking water, adequate sewage disposal, and immunization against childhood diseases. But the theme of a recent WHO's World Health Day, sponsored by WHO and celebrated each year on April 7, was "Move for Health," reflecting the fact that non-communicable diseases have become epidemic due largely to a worldwide shift in lifestyles leading to reduced physical activity, changing diets, and more tobacco use. What else do you know about the health of the world? Open wide for this quick quiz.

1. **In 1979, WHO announced the eradication of what major killing disease?**
2. **What is the next likely disease to be eradicated?**
3. **According to WHO, what is the most devastating disease humankind has ever faced?**
4. **What worldwide problem kills at least 3 million children under age 5 each year?**
5. **True or False? By WHO estimates, nearly half a million women died of "maternal causes" in 2005.**

ANSWERS

1. Smallpox, one of history's deadliest killers. The last known naturally occurring case was in Somalia in 1977 and the last known case was due to a lab accident in England in 1978. Two research stores of smallpox are secured in labs in the United States and Russia.

2. A polio-free world is within grasp. The number of polio cases has dropped by 99 percent, and the world soon may see the last of a disease that has killed and crippled millions of children throughout the ages.

3. AIDS. According to United Nations statistics, since the epidemic began in 1981, more than 25 million people have died of AIDS. In 2006, an estimated 39.5 million people were living with HIV (some estimates go as high as 47.1 million), with 64 percent living in sub-Saharan Africa where there are some 12 million AIDS orphans. However, according to the U.N., infection rates were rising in East Asia, Eastern Europe, and Central Asia. There were more than 3 million estimated deaths from AIDS in 2006. Worldwide, it is the fourth biggest killer. HIV/AIDS is now the leading cause of death in sub-Saharan Africa.

4. Environmental hazards. According to WHO, inadequate drinking water and sanitation, indoor air pollution, and accidents, injuries, and poisonings cause approximately 3 million deaths suffered annually by children under age 5.

5. True. "Maternal mortality," defined as death during pregnancy or within forty-two days of delivery: 99 percent of those deaths were in developing countries, and more than half occurred in Sub-Saharan Africa.

EVERYDAY OBJECTS AND REMARKABLE INVENTIONS

DON'T KNOW MUCH ABOUT *The Mustang*

WHILE AMERICANS ARE KNOWN to love their cars, sometimes the relationship goes a step beyond. Only a few cars make that grade—like the T-Bird, 'Vette, and, of course, the Mustang. Introduced in 1964, the Mustang was an instant hit. Its price and style made it a record-breaking competitor to GM's Corvette. After a blitz of advertising, frenzied customers bought 22,000 Mustangs on the first day and a remarkable million sold within two years. Maybe "All you want to do is ride," but "slow your Mustang down" for this quick quiz.

1. **What notable auto industry executive was behind the Mustang?**
2. **At what event was the Mustang officially introduced?**
3. **In what film did the Mustang first appear?**
4. **How much was a Mustang in 1964?**
5. **What pop diva came up with the lyrics "Ride Sally Ride" in the hit song "Mustang Sally"?**

ANSWERS

1. Lee Iacocca headed the team of engineers and designers who developed the Mustang. He joined Ford as an engineer but switched to sales. From 1970 to 1978, Iacocca served as president of Ford. Later, as chairman of the board of the Chrysler Corporation, Iacocca helped save the company from bankruptcy.
2. On the drawing board since 1962, it debuted publicly on April 14, 1964, at the World's Fair. It went on sale on April 17, 1964.
3. A yellow Mustang hit the screens in *Goldfinger*. James Bond gave chase in his famed Aston Martin.
4. At $2,320.96, it was about half the price of a Corvette.
5. Aretha Franklin suggested the lyrics to composer Mack Rice whose original title was "Mustang Mama." The song became a big hit for Wilson Pickett.

DON'T KNOW MUCH ABOUT *Sports Cars*

IN 1903, AMERICA'S love affair with the car began with Ford Motors. Fifty years later, the affair got torrid. On June 30, 1953, the first Corvette rolled off a General Motors assembly line. America had its first sports car. Ford's Model T's had been durable and cheap. The Corvette was a two-seater with a low, streamlined body built for speed and performance. Even today, the romance with the Corvette has not cooled and the car has become an icon of American style. What do you know about sports cars? Start your engines for this quick quiz.

1. **What famed auto designer, featured in a series of automobile ads, was behind the Corvette's introduction?**
2. **What was the price tag on a 1953 Corvette?**
3. **What car supposedly inspired the Corvette?**
4. **What was Ford's response to the Corvette?**
5. **The year before the 'Vette debuted, GM chief executive Charles Wilson made a memorable comment to a Senate committee. What did he say?**

ANSWERS

1. Harley Earl, then head of GM's Special Projects crew.
2. A little more than $3,000.
3. The Jaguar XK120 is believed to have been the inspiration.
4. In 1954, Ford answered with another American classic, the Thunderbird, which was later turned into a four-seater sedan.
5. Wilson told the Senate Committee on Armed Services, "For years I thought what was good for our country was good for General Motors, and vice versa. The difference did not exist." Wilson's remark is often misquoted. He later became Eisenhower's Secretary of Defense from 1953 to 1957.

DON'T KNOW MUCH ABOUT *Trains*

"ALL ABOARD!" THE conductor's call still has a ring of romance. Before Americans fell in love with the auto, they loved the train. But since the 1940s, passenger trains fought a losing battle against airplane and automobile. On May 1, 1971, Congress created Amtrak, a private company, in an attempt to revive passenger rail service. Today, the nationwide rail system consists of 22,000 miles of rails linking 500 communities in forty-six states. And as frustration with air and highway travel grows, the train is making a comeback. In 2006, Amtrak served 24.3 million passengers, the fourth straight year of record ridership. America celebrates a memorable date in railroad history on May 10, the day the famed Golden Spike was driven into place, linking east and west by rail in 1869. What do you know about America's railroading past? Climb aboard for this quick quiz.

1. **Where did eastern and western railroads have their historic meeting in 1869?**
2. **What became of the famed Golden Spike?**
3. **Which President signed the bill calling for a transcontinental railroad?**
4. **To what famous battle did troops arrive by railroad for the first time?**
5. **How long did it take to cross America by train in 1869?**
6. **How fast does the new generation Acela train travel?**
7. **What are the top five busiest train stations?**

ANSWERS

1. At Promontory Point, Utah, on May 10, 1869.
2. Five ceremonial spikes were actually used. All were removed and replaced by ordinary spikes driven in by a Chinese laborer.
3. Abraham Lincoln in 1862.
4. The Battle of Bull Run or First Manassas in 1861.
5. Coast to coast in ten days.
6. At speeds up to 150 mph, the Acela now makes the New York to Washington trip in less than two and a half hours.
7. In 2006, they were: New York, Washington, Philadelphia, Chicago, and Los Angeles.

DON'T KNOW MUCH ABOUT *ENIAC*

IT SURVIVED Y2K. But how will your computer manage its midlife crisis? That's right. The modern electronic computer has hit its sixties. Although preceded by earlier specialized computers, the first all-electronic, all-purpose computer, fondly known as ENIAC, was unveiled at the University of Pennsylvania on February 14, 1946. It might have been Valentine's Day, but it was not love at first sight. ENIAC was a long way from today's laptop or handheld computer. Weighing in at thirty tons and containing 18,000 vacuum tubes, it was about the size of a motor home and drew so much electricity that lights in a nearby town dimmed each time it was used. Can't live without your PC? What do you know about its past?

TRUE OR FALSE?

1. **The name ENIAC was the result of a lab joke: Every Night It Alters Currents.**
2. **ENIAC was designed to solve the mathematical puzzle known as Fermat's Theorem.**
3. **The first electronic calculating device was actually produced in England before ENIAC.**
4. **In 1950, ENIAC was used to make the first computerized weather predictions.**
5. **ENIAC's patent was invalidated in 1973.**

Answers

1. False. ENIAC stands for Electronic Numerical Integrator And Computer.
2. False. It was created during World War II because the Ballistic Research Laboratory in Aberdeen, Maryland, needed a faster computer to calculate trajectories for firing tables used by gunners.
3. True. The "Colossus" machine, created by a team led by mathematician Alan Turing, was devoted to cracking the German secret code known as Enigma.
4. True, ushering in a whole new age of complaints about weather forecasts.
5. True. John Vincent Atanasoff was then credited as the "inventor" of the modern computer.

DON'T KNOW MUCH ABOUT
American Fashion

FOR A LONG TIME, the fashion world was dominated by the famous houses of Europe who looked askance at Americans. But in recent years, American designers and companies have come to the fore in the trendsetting world of clothing. What do you know about some high points in American fashion? Try on this quick quiz.

1. **What fashion icon was created in 1967 by Bronx-born Ralph Lifshitz?**
2. **What fashion sensation did First Lady Jackie Kennedy create on Inauguration Day 1960?**
3. **Begun in 1921 in Florence, Italy, as a saddlery shop, what legendary European company was run by American Tom Ford for ten years?**
4. **What clothing item did the word "shoddy" first describe?**
5. **What American fashion revolution began in 1850 with the bibless overalls created by a 20-year-old Bavarian-born entrepreneur?**
6. **New York shirtmaker Oliver Winchester was better known for what product?**

ANSWERS

1. The Polo brand was born that year with a line of ties designed by the man better known as Ralph Lauren.

2. Her husband wore a top hat, but Mrs. Kennedy's pillbox hat became a signature fashion item of the 1960s.

3. Gucci, founded by leather designer Guido Gucci whose initials form the company's signature logo.

4. In the Civil War, Union uniforms were made from this cheap cloth that literally unraveled, giving the word its modern meaning.

5. That was when Levi Strauss began the company that would eventually create the blue jean.

6. His shirtmaking success allowed him to buy Volcanic Repeating Arms, which was later named the Winchester Repeating Arms Company, manufacturer of the famous rifle.

DON'T KNOW MUCH ABOUT *Fashion*

LED BY A PARADE of new Easter dresses and bonnets, Spring is often hailed as a time for new fashion. But of course, there is nothing new about fashion, which has always played a role in our past. Are you a fashion dropout? Slip into this quick quiz.

1. **For what uplifting fashion is American Mary Phelps Jacob famed?**
2. **In colonial America, "Breeching Day" was a significant moment. What did it mean?**
3. **What notable fashion was inspired by a 1946 atomic bomb test?**
4. **In the 1850s, an early feminist leader added what word to the fashion lexicon?**
5. **What fashion indicator is said to predict recessions and recoveries?**
6. **John Jacob Astor parlayed what fashion rage into one of the first great self-made American fortunes?**

Answers

1. A descendant of steamboat pioneer Robert Fulton, Mary Jacob designed the first elastic brassiere out of handkerchiefs and pink ribbon, an idea she patented in 1914, although the bra concept dates to Roman times. Jacob later sold the patent to the Warner Brothers Company, which introduced cup sizing in 1939. Another claim is staked by Parisian dress designer Philippe de Brassiere who pioneered the "chest halters" to which his name was later attached.

2. This was the day that a young boy gave up his childish short pants for the longer "breeches" worn by men, source of the slang word "britches" for pants.

3. The bikini was named following a French test of two atomic bombs at the Bikini Atoll in the Pacific Ocean. The revealing swimsuit made its debut soon after on French beaches and the name caught on, perhaps because the suit had such an explosive effect on men. Others suggest it had more to do with the twin mushroom clouds that the tests produced.

4. Bloomers. Although she did not invent them, suffragist leader Amelia Jenks Bloomer helped to popularize the billowing undergarment covered by a skirt—the nineteenth-century version of a pantsuit for women.

5. According to fashion sources, hemlines on skirts go down in recessions and rise with an economic recovery. Lipstick sales are also said to increase in recessions as women look for an inexpensive way to change their looks.

6. A former baker boy and peddler, Astor traded flannel and "firewater" with Indians for animal pelts, especially beaver for men's hats. Astor converted his fur riches into large real estate holdings in Manhattan.

SPACE AND THE NATURAL WORLD

DON'T KNOW MUCH ABOUT *Mosquitoes*

JOKINGLY REFERRED TO AS "The Texas State Bird," the mosquito is an excuse for three days of fun and games at the Great Texas Mosquito Festival held in Clute, Texas. Featuring mascot "Willie Man-chew, the World's Largest Mosquito," this celebration has gained national recognition. But there is nothing fun about this insect that is responsible for spreading some of the world's deadliest diseases, including malaria, encephalitis, and West Nile Virus. What do you know about these pests? Get the buzz on mosquitoes in this quick quiz.

1. **What does mosquito mean?**
2. **Which mosquitoes cause the problems, male or female?**
3. **Do mosquitoes "bite" or "sting"?**
4. **Where can you go to avoid them completely?**
5. **Who do you call if you want a good "mosquito-buster"?**

ANSWERS

1. It is derived from the Spanish word for "little fly."
2. Only a female mosquito will "bite," as they require blood to produce their eggs. The male prefers the taste of flower nectar.
3. Neither. A female mosquito actually siphons the blood out by a tube.
4. Antarctica is the only place mosquitoes do not live.
5. Bats and dragonflies. While a single dragonfly can eat up to 600 insects a day, a brown bat can put away 1,200 mosquitoes in an hour!

DON'T KNOW MUCH ABOUT *Infectious Diseases*

The aftermath of Hurricane Katrina offered a reminder of the dangers posed by infectious diseases. Throughout history, such diseases as smallpox and AIDS have wiped out far more people than wars. Held each year in the third week in October, International Infection Prevention Week is marked in an attempt to raise awareness about the basic preventive measures—hand washing, clean water, insect repellent—that can be taken to guard against diseases, both at home and while traveling. What do you know about the dangers of infectious diseases?

1. **What is cholera?**
2. **Is smallpox still a danger?**
3. **What is "Bird Flu"?**
4. **What is West Nile Virus?**

ANSWERS

1. An acute intestinal infection caused by ingesting bacteria-contaminated water. As a rule for prevention, the Center for Disease Control advises: "Boil it, cook it, peel it. Or forget it."

2. One of the most deadly killers in history, smallpox is a viral disease that was declared eradicated in 1980. The threat of a biological weapon makes the need for smallpox preparedness a reality. In the event of a smallpox outbreak, there is now enough vaccine for every person in the United States, says the CDC.

3. Bird flu (avian influenza) is caused by a virus occurring naturally in birds. According to the CDC, the risk to humans, especially in America, is low. The CDC has advised travelers to parts of Asia to avoid poultry farms, contact with live animals in markets, and surfaces that might be contaminated with animal waste. (The CDC maintains a Web site: www.cdc.gov.)

4. West Nile Virus is a potentially serious illness, spread largely by infected mosquitoes. (A small number of cases have been through transfusions, transplants, and breast feeding.) Prevention tips include: use of an EPA-approved insect repellent and elimination of mosquito-breeding sites, such as buckets, wading pools, bird baths, and other places in which water collects.

DON'T KNOW MUCH ABOUT *The Metric System*

AMERICANS DON'T LIKE TO give an inch—especially to things French. That's been true when it comes to metric measurements, based on the decimal system. Formally known as the International System of Units (or SI, the initials of its French name), the modern metric system is more logical, precise, and internationally understood than the American inches and pounds system. Americans are slow to accept these worldwide standards. Still don't know a liter from a meter? See how you measure up in this quick quiz.

1. **Where and when did it originate?**
2. **What countries don't use it?**
3. **Who proposed the first metric system in America?**
4. **What was the first decimal-based currency?**
5. **What is a meter based on?**
6. **What was the U.S. Metric Board?**

ANSWERS

1. France proposed an international measurement system in 1790 and officially adopted the metric system in 1795. By 1900, most of the commercially advanced countries of the world had adopted it.

2. For years, the main holdouts were the United States and nations of the British Empire. The UK, Canada, Australia, and New Zealand have largely completed the switch, and almost all official U.S. government, scientific, technical, and commercial interests use the metric system. According to the U.S. Metric Association, the only other known holdouts are Liberia and Myanmar (formerly Burma).

3. Thomas Jefferson in 1790. John Quincy Adams also proposed the switch but Congress rejected the idea.

4. The U.S. dollar based on 100 cents, created in 1792.

5. Originally a meter was calculated as one-ten-millionth of the distance from the Equator to the North Pole. In 1983, it was recalculated in terms of the speed of light traveling in a vacuum.

6. Formed by an act of Congress in 1975 to coordinate the increasing conversion to metrics, the board was disbanded by President Reagan in 1980. Today most U.S. government agencies use metric measurements.

DON'T KNOW MUCH ABOUT *The Pacific Ocean*

IT IS BIG. Really big! Larger than all the land in the world put together, the Pacific Ocean covers about 70 million square miles—almost one third of the Earth's surface. September 26, 1513, is traditionally marked as the day the Pacific was "discovered" by Spanish explorer Vasco Nuñez de Balboa. But in fact, he was simply the first European to see the Eastern Pacific. People had sailed the Pacific Ocean for thousands of years before Balboa saw it. What else do you know about the world's largest body of water? Dive into the deep blue with this quick quiz.

1. **Balboa found it for Spain and called it El Mar de Sur (South Sea). Who named it "Pacific"?**
2. **What is the "Ring of Fire"?**
3. **Where is the deepest known spot in any ocean?**
4. **True or False? The Pacific contains more than 20,000 islands.**

ANSWERS

1. Portuguese explorer Ferdinand Magellan, sailing for Spain, named it with the Spanish word for "peaceful" in 1520. The first European to sail across the Pacific during his attempt to sail around the world, Magellan died in 1521 in a fight with natives of the islands the explorer named the Philippines.
2. A string of about 300 active volcanoes that rims the Pacific, going from the tip of South America, north to Alaska, then west to Asia through Japan, the Philippines, Indonesia, and New Zealand.
3. The Challenger Deep in the Mariana Trench, near Guam, extends 36,198 feet below the surface. The average depth of the Pacific is almost 13,000 feet and its deepest parts are actually near the shores.
4. True. The Pacific contains 20,000 to 30,000 islands; the exact number has not been determined.

DON'T KNOW MUCH ABOUT *Cold Weather*

THEY WON'T EASILY FORGET January 23 in Prospect Creek, Alaska. It was there that the record for the lowest temperature ever measured in the United States was set in 1971; the mercury bottomed out at a chilly -80°F. (A Yukon, Canada, airport called Snag holds the North American record at -81°F.) Of course, while some might be feeling Old Man Winter's grip, the normal January high in Honolulu is 80°F. Feeling the winter blues? What do you know about cold weather? Curl up by the fire with this quick quiz.

1. **Where was the world's coldest temperature recorded?**
2. **What is the world's coldest inhabited place?**
3. **When did the last Ice Age end?**
4. **What is the "sublimation of water vapor into solid crystals at temperatures below freezing"?**
5. **Which wintry named poet wrote the famous verse, "Stopping by Woods on a Snowy Evening"?**

ANSWERS

1. The Soviet station at Vostok in Antarctica set the all-time low with a reading of -129°F in July 1983.
2. Norilsk, Russia, where the average temperature is 12.4°F.
3. According to the Geologic Time Scale, the last of the great Ice Ages, which ran in stages during the Pleistocene epoch, ended about 11,000 years ago.
4. Snow.
5. Robert Frost.

DON'T KNOW MUCH ABOUT *Tulips*

PASADENA HAS ITS ROSE BOWL. But Holland has Tulip Time. A ten-day festival in Holland, Michigan, Tulip Time attracts a half million visitors who flock to America's Tulip Capital for this springtime event, which sometimes peaks on Mother's Day. What do you know about the past of this popular spring flower? Try this quick quiz.

TRUE OR FALSE?

1. **Tulips originated in the Netherlands.**
2. **A crash in the tulip market bankrupted hundreds of Dutchmen in 1637.**
3. **Tulips were brought to America by Pilgrims.**
4. **In the 1960s, the Beatles reintroduced the old favorite, "Tiptoe Through the Tulips."**

Answers

1. False. The first tulip bulbs were brought to Europe from Turkey in the 1500s. They became a sensation in Holland.
2. True. Between 1634 and 1637, "Tulipomania" gripped the country and speculation drove the price of tulips up dramatically, with bulbs worth far more than their weight in gold. The speculative bubble burst and the government began to regulate the industry.
3. False. Tulips were introduced into America by Dutch settlers in New Amsterdam, later New York.
4. False. The singer named Tiny Tim, playing his ukulele, made the song a big hit.

DON'T KNOW MUCH ABOUT *Mount St. Helens*

ON MAY 18, 1980, the Mount St. Helens volcano in the state of Washington blew its top in an extraordinary explosion. The most destructive eruption in U.S. history, it left more than sixty people dead. What do you know about that blast?

TRUE OR FALSE?

1. Mount St. Helens is now taller than before the eruption because of massive lava deposits.

2. Mount St. Helens is now an extinct volcano.

3. Volcanic ash from that 1980 explosion traveled more than fifteen miles into the air.

4. There are more than forty major active volcanoes in the United States.

ANSWERS

1. False. Before May 18, 1980, Mount St. Helens was 9,677 feet tall; it is now 8,364 feet.
2. False. Mount St. Helens remains a potentially active and dangerous volcano.
3. True. The ash, composed of pulverized rock and new lava, was carried eastward across the country. Trace amounts of ash carried around the world.
4. True. There are 90 in Alaska alone, but there are also potentially active volcanoes in California, Oregon, Washington, Hawaii, Arizona, and New Mexico.

DON'T KNOW MUCH ABOUT *Summertime*

As the classic Gershwin tune put it, "Fish are jumping and the cotton is high." While most of us know that summertime officially arrives on June 21 with the summer solstice, it's still a hot topic. What do you know about summertime heat?

True or False?

1. It gets hotter in summer because the Earth is closer to the Sun.

2. Summer in North America is winter in South America.

3. The highest temperature recorded on earth was in Death Valley, California.

4. The scientific name for the Sun is Solarium Centrum.

Bonus: What famous American general said, "I propose to fight it out on this line if it takes all summer"?

Answers

1. False. During our summer, the Earth is actually more distant from the Sun. But the Northern Hemisphere is tilted on the Earth's axis toward the Sun, giving it more direct sunshine.

2. True. While the Northern Hemisphere tilts toward the Sun, the Southern Hemisphere is angled away, reversing the seasons.

3. False. The planet's highest temperature was 136°F, recorded in Libya; the U.S. record high is 134°F in Death Valley.

4. False. The Sun is simply called the Sun.

Bonus: General Ulysses. S. Grant in 1864.

DON'T KNOW MUCH ABOUT *The Aurora Borealis*

THE CHINESE CALLED IT "candle dragon," the ancient Romans "blood rain." Most people know it as the aurora borealis, or "Northern Lights" (the literal translation is "northern dawn"). It is one of nature's greatest spectacles. The phenomenon often peaks in the summertime, with the auroral light show sometimes reaching down as far as the southern United States. How "bright" are you about this natural spectacle?

TRUE OR FALSE?

1. **The aurora is caused by sunlight filtering through atmospheric dust.**
2. **You can't see the aurora in the Southern Hemisphere.**
3. **The aurora is a key indicator of what scientists call "space weather."**
4. **Auroral storms are beautiful but not destructive.**

ANSWERS

1. False. The aurora is the result of "solar wind," the stream of ions and electrons emanating from the Sun, which collide with oxygen and nitrogen in the atmosphere high above the Earth's magnetic poles. This combination produces the brilliant spectrum of colors in the aurora.

2. False. In the Southern Hemisphere it is called "aurora australis."

3. True. Space weather—the interaction between the Sun and the Earth—is created by the solar wind that streams constantly out from the Sun. Powerful flares can whip that solar wind to dangerously high levels. The Sun's turbulence runs in eleven-year cycles.

4. False. A powerful auroral storm in 1989 knocked out Quebec's electricity for nine hours. The high-energy burst of solar winds can also disturb radio communications, satellites, radar systems, and other sophisticated electronics on which we increasingly rely.

DON'T KNOW MUCH ABOUT *Diamonds*

THEY ARE, OF COURSE, said to be "forever" and "a girl's best friend." They are just lumps of coal that time, heat, and pressure turn into the earth's hardest natural substance. Natural diamonds form beneath the earth's crust, where high temperatures and pressure cause the diamonds to crystallize. The largest stone ever discovered was found a hundred years ago in South Africa: the Cullinan weighed 3,106 carats (about 1⅓ pounds). What do you know about this much-prized "ice"? Polish off this quick quiz.

1. **What is the "Star of Africa"?**
2. **If it's so hard, how do you cut a diamond?**
3. **What are the "Four C's" of diamonds?**
4. **Where are the most diamonds mined?**
5. **What is a "conflict diamond"?**

ANSWERS

1. The Cullinan was cut into into nine large gems and ninety-six smaller stones. The largest of these is the 530-carat "Star of Africa," among the Crown Jewels of England.

2. A diamond can only be cut by another diamond, but a diamond can be cleanly broken with a sharp, accurate blow because of "cleavage," the ability to split in certain directions.

3. Gem diamonds are graded according to Carat (weight), Clarity, Color, and Cut.

4. Australia outranks all other countries. The United States has no commercial diamond mines. However, kimberlite, a volcanic stone that contains diamonds, has been found in Arkansas and a few other states.

5. Conflict diamonds are diamonds that originate in areas controlled by rebels and other factions opposed to legitimate governments, and are a crucial factor in prolonging wars in parts of Africa. In 2000, the United Nations General Assembly adopted a resolution trying to break the link between the illicit transaction of rough diamonds and armed conflict.

SPORTS

DON'T KNOW MUCH ABOUT *The Iditarod Sled Dog Race*

THE RUNNING OF the Iditarod Sled Dog Race mushes off each year in late winter from downtown Anchorage. In what is called the "Last Great Race on Earth," more than eighty teams of dogs and drivers head for Nome, Alaska, confronting arctic temperatures, rugged terrain, and other challenges posed by the beautiful but harsh Alaskan wilderness. The race is not without controversy, as animal rights activists criticize it. But supporters argue that sled dogs are prized, well-treated animals whose welfare is a priority, and veterinarians check them along the route. What do you know about this daunting challenge to the men, women, and dogs who take on the "more than 1,049-mile" race?

TRUE OR FALSE?

1. **The route of the race is actually 1,200 miles long.**
2. **The sled race commemorates the rush to the Yukon gold fields.**
3. **The most famous sled dog of all, Balto, is honored by a statue in Washington, D.C.**
4. **The last finisher receives a traditional Red Lantern.**
5. **"Mush" is derived from the French *marchons*.**

ANSWERS

1. True. The figure "1,049" is symbolic of a thousand mile race in the forty-ninth state.

2. False. In 1925, relay teams of dogs carried diphtheria serum from Anchorage to Nome to prevent an epidemic.

3. False. The famous statue of Balto is in New York's Central Park. Although Balto was the lead dog on the team that delivered the serum, another famed dog, Togo, led his team for an astonishing 260 miles.

4 True. It is a symbol of the days when lanterns were hung on roadhouses when "mushers" were out on the trail and were not extinguished until they returned safely.

5. True. The word for "let's go" or "hurry up" used by French trappers, it was gradually transformed into "Mush on!" and eventually just "Mush!"

DON'T KNOW MUCH ABOUT *Pro Hockey*

FANS WEATHERED A LABOR DISPUTE that shut down the 2004–2005 National Hockey League season. There's not much protection against a strike or lockout. But hockey players of the modern era do get some protection on the ice. Tired of taking pucks in the face—and the stitches and lost teeth that go with them—a goalie donned a protective mask in November 1945 and changed the face of hockey, literally. Imagine what Jason of *Friday the 13th* fame would be without his hockey mask!

1. **Who wore the first mask in National Hockey League history?**
2. **What were the first teams in the National Hockey League?**
3. **What was the first American team in the NHL?**
4. **What player has the most goals in a season? In a career?**
5. **Who is "Lord Stanley"?**
6. **What came first: field hockey or ice hockey?**

ANSWERS

1. Jacques Plante of the Montreal Canadiens is credited with starting the trend with a fiberglass mask; Clint Benedict had experimented with a leather mask in 1930, but the idea never caught on.

2. The league formed in 1917 with the Montreal Canadiens, Montreal Wanderers, Ottawa Senators, and Toronto Arenas.

3. The Boston Bruins who joined in 1924. Chicago, Detroit, and Pittsburgh and two teams from New York City joined in 1925 and 1926.

4. One answer: The Great One, Wayne Gretzky, who holds the records for goals in a single season (92) and regular season goals in a career (894), among many other records.

5. The NHL's championship trophy is named for Baron Stanley of Preston, Governor General of Canada, who donated a silver cup to go to Canada's champion amateur hockey team in 1893. The names of the winning teams and players are engraved on the cup.

6. Field hockey. The first formal hockey game supposedly took place in Ontario, Canada, in 1855, although games may have been played earlier in Nova Scotia. The official rules were drawn up at Montreal's McGill University in the 1870s.

DON'T KNOW MUCH ABOUT *Baseball Parks*

Green grass. Green Monster. Green ivy. The names are legends of Americana: Fenway, Wrigley and, of course, Yankee Stadium. To purists, these are real ballparks and the newer stadiums—PNC Park, Pacific Bell Park, Network Associates Coliseum—sound more like a stock portfolio than a baseball lover's dream. But that is modern baseball. What else do you know about the homes of America's beloved baseball? Step up to the plate for this quick quiz.

1. **Which current stadium is the oldest?**
2. **Which is the newest stadium?**
3. **Which stadium has the largest capacity?**
4. **Only three stadiums still use artificial turf. Which ones?**
5. **Beer leads juice when it comes to naming stadiums. Which are named for beers and which for juice makers?**
6. **What happened to Houston's Enron Field?**
7. **What was the last park to get lights installed?**

ANSWERS

1. Boston's Fenway, home to the legendary Green Monster wall, was built in 1912.

2. Busch Stadium, home of the St. Louis Cardinals, opened in April 2006. But it was surpassed when Nationals Park opened in Washington, D.C., in 2008. But over the next few years, new stadiums are slated to open in several cities. In New York, both the Mets and Yanks will have new homes in 2009. Minnesota (2010), Oakland (2011), Miami (2011), and Tampa Bay (2012) all have parks planned or under construction.

3. In 2008, the major league baseball capacity champion was the legendary Yankee Stadium, the "House That Ruth Built" in the Bronx, with a capacity of 56,937. Construction of Yankee Stadium II is underway, and the new park is scheduled to open in 2009 with a smaller capacity of 51,800. Once completed, Yankee Stadium II will surrender the capacity crowd to Dodgers Stadium in Los Angeles, which holds 56,000 fans.

4. Only three teams used the hated grass replacement as of the 2007 season: the Minnesota Twins, Tampa Bay Devil Rays, and Toronto Blue Jays, all in the American League.

5. Coors (Rockies), Busch (Cardinals), and Miller (Brewers) are named for beer makers. Tropicana (Devil Rays) and Minute Maid (Astros) are for juice companies.

6. After the company's collapse, the Houston stadium was renamed Minute Maid Park.

7. Venerable Wrigley in Chicago, built in 1914, where night baseball was played for the first time in 1988.

DON'T KNOW MUCH ABOUT *The Soap Box Derby*

It is a scene right out of a Norman Rockwell painting. Happy boys speeding downhill in their homemade contraptions in the "Gravity Grand Prix." The Depression-era creation of an Ohio news photographer who saw some boys racing their simple crate-and-buggy-wheel cars, the All American Soap Box has become an international event since its beginning in 1933 in Akron, Ohio. What do you know about this slice of American pie? Try this quick quiz .

True or False?

1. **In the early years, the rules demanded the use of a soap box to build the racers.**
2. **The official symbol of the race, "Old No. 7," never won the Derby.**
3. **The event has run without interruption since 1933.**
4. **The tradition of "boy-built cars" remains; girls cannot compete.**
5. **Under the rules, contestants must build their own racers.**
6. **In a 1951 celebrity version, the runner-up was actor Ronald Reagan.**

Answers

1. False. "Orange Crate" derby would have been more accurate. There is no record of anyone ever using a soap box to build a "soap box" racer.
2. True. The photographer who conceived the race took a picture of "Old No. 7" because he thought it looked like the ideal soap box racer, but it did not win a race.
3. False. After Pearl Harbor was bombed, there were no races for the next four years.
4. False. Females were allowed to participate beginning in 1971.
5. True. Judges check out the contestants, who sometimes must prove they are capable of building the racers. In one notorious race, the 1973 winner was discovered to have placed an electromagnet in the front of his racer.
6. True. Reagan finished behind ventriloquist Paul Winchell and his dummy, Jerry Mahoney.

DON'T KNOW MUCH ABOUT *Olympic Moments*

WHEN THEODOSIUS I, an early Christian Emperor of Rome, banned the "pagan" Olympics in 393 AD, it marked the first but not last time that politicians spoiled the sports. The Olympic Games originated in Greece where the first recorded games took place in 776 BC at Olympia. Part religious festival, the games were so important that Greeks measured time in Olympiads, the four-year interval between games. Ah, for the good old days! Things just haven't been the same since Kyniska of Sparta took the four-horse chariot race in 396 BC, and became the first woman Olympic champion. Since the revival of the Games in 1896, there is no longer a race in armor, chariots pulled by mules, or trumpet competition. What hasn't changed is that politics still often intrudes on the Games. What do you know about these Olympic moments? Go for the gold with this quiz.

1. **Which hero of the 1912 Stockholm Games had his medals taken away on a charge of "professionalism"?**
2. **Which Olympics were canceled because of wars?**
3. **Who won four gold medals in 1936 Berlin, discrediting Hitler's racist propaganda?**
4. **At which Games were two medal-winning Americans suspended by the Olympic Committee after they raised clenched fists at a ceremony?**
5. **What tragedy occurred at the 1972 Munich Games?**

ANSWERS

1. American Indian Jim Thorpe, who had received money for playing baseball. His medals and place in the records were restored in 1982.

2. 1916 (World War I); 1940 and 1944 (World War II).

3. American track star Jesse Owens.

4. 1968 in Mexico City. The runners Tommie Smith and John Carlos raised their fists and bowed their heads to protest racism.

5. Palestinian terrorists killed eleven Israeli athletes. The terrorists and a German policeman also died in a shootout.

DON'T KNOW MUCH ABOUT *Football*

THE REST OF THE WORLD has its "football," which Americans call soccer. But America has football. Borrowing from England's soccer and rugby, the American game emerged in the late nineteenth century. As fast and violent as it is today, football was even more rugged back then. After eighteen players died in 1905, President Teddy Roosevelt called for rule changes in order to prevent banning the game entirely. The National Collegiate Athletic Association (NCAA) was formed in 1908 to set rules for the sport. Are you ready for some football? Kick off with this quick quiz.

TRUE OR FALSE?

1. **The first forward pass was thrown by Knute Rockne in the 1913 Notre Dame–Army game.**

2. **The "pigskin" was never really made from a pig's skin.**

3. **The first collegiate football game was played between Harvard and Yale.**

4. **Vince Lombardi never said, "Winning isn't everything. It's the only thing."**

Answers

1. False. According to historians of the game, the first forward pass was thrown in a 1906 contest between Wesleyan and Yale. In the 1913 season, Notre Dame receiver Knute Rockne (later Notre Dame's great coach) and quarterback Gus Dorais made the forward pass a primary weapon, revolutionizing the game.

2. True. Though animal bladders were once booted around as makeshift balls, the carcass of a pig was never used to make footballs, which have always been covered in leather. The term dates to 1894, when it was used by Amos Stagg's University of Chicago Maroons, a powerhouse team of that era.

3. False. The first recognized collegiate game was played between Princeton and Rutgers on November 6, 1869.

4. False. Although the Green Bay Packers' legendary Coach Lombardi later protested that he had not said it, the now-famous quote appeared in several forms. But Lombardi was not the first to say it. That honor goes to Vanderbilt coach Henry Sanders who was so quoted in a 1955 *Sports Illustrated*.

DON'T KNOW MUCH ABOUT *The Kentucky Derby*

RUN EACH SPRING, the Kentucky Derby is the oldest continually held sporting event in America. How much do you know about America's most famous horse race and its old Kentucky home?

TRUE OR FALSE?

1. The Derby is named after the famous hats, once worn by jockeys.
2. The length of the race has always been one and a half miles.
3. Kentucky was the fifteenth state.
4. Louisville, home of Churchill Downs, is the state capital.
5. The Derby is the first leg of racing's Triple Crown. The others are the Preakness and Saratoga Stakes.

ANSWERS

1. False. Both the race and the hat are named after Edward Stanley, the Earl of Derby, who founded the English Derby, the model for the Kentucky Derby.
2. False. Originally that distance, the race was later reduced to the current mile and one-quarter.
3. True. Once part of Virginia, Kentucky entered the Union in 1792. Although a slave state, it remained in the Union during the Civil War.
4. False. Frankfort is the state capital.
5. False. The third leg is the Belmont Stakes, not Saratoga.

DON'T KNOW MUCH ABOUT *Baseball Geography*

Most of us know that Abner Doubleday did not invent baseball. Derived from the British games of cricket and rounders, varieties of baseball had been played since colonial times until New Yorker Alexander Cartwright set down the rules and the first "official" game was played in June 1846 in Hoboken, New Jersey. Since then, baseball has grown into the "national pastime" with teams sprouting up all over the country—and eventually into Canada and now around the world. Baseball teams, like the beloved Dodgers and Giants of New York, have moved with the times, usually leaving fans brokenhearted. Where do these current baseball teams get their roots? Test your baseball geography.

1. **The Atlanta Braves once hailed from Milwaukee. But in which eastern city did the Braves originate?**
2. **They are Baltimore's Orioles today, but once they were the Browns of what river city?**
3. **They are Twins in Minnesota but once adjourned in another district.**
4. **They left the heartland and became the Oakland Athletics.**

ANSWERS

1. Boston. The National League team moved to Milwaukee in 1953 and Atlanta in 1966.

2. The St. Louis Browns moved to Baltimore in 1954.

3. The Washington Senators moved to Minnesota in 1961. A new Senators franchise later left Washington and became the Texas Rangers.

4. The Kansas City Athletics moved west in 1968.

DON'T KNOW MUCH ABOUT *The Ivy League*

THEY WEREN'T PLAYING football when King's College was founded in colonial New York in 1754. In postrevolutionary America, the name was changed to Columbia in 1784. Fast forward to 1933 when a New York sportswriter coined the term "Ivy colleges" for these eight prestigious schools—among the nation's oldest—with their distinctive ivy-covered walls and buildings, which pioneered college football. The name stuck with sportswriters, and the football alliance expanded to an official league including all intercollegiate sports in 1954. Today the Ivy League, better known for its brains than athletic prowess, still fields teams in many sports, and boasts several national champions. What else do you know about the Ivy League? Put on your thinking cap for this quick quiz.

1. **What are the eight members of the Ivy League?**
2. **Which is the oldest member? Which is the newest?**
3. **Which Ivy League school is the oldest university?**
4. **Which of these colleges was begun "for the education and instruction of Youth of the Indian Tribes in this Land"?**
5. **How many Ivy League university presidents have become the U.S. President?**
6. **Which Ivy League Rhodes Scholar starred in pro basketball and politics?**

Answers

1. (In alphabetical order) Brown, Columbia, Cornell, Dartmouth, Harvard, University of Pennsylvania, Princeton, and Yale. In the early days of the Ivy League, Army was also included.
2. Harvard, founded in 1636, is the oldest secondary school in America. Cornell, located in Ithaca, New York, was founded in 1865.
3. Philadelphia's University of Pennsylvania (aka Penn, but not to be confused with Penn State) was founded as a university by Benjamin Franklin in 1749; Harvard College became a university in 1790.
4. Dartmouth, in Hanover, New Hampshire, founded in 1769.
5. Two: Woodrow Wilson (Princeton) and Dwight Eisenhower (Columbia).
6. Princeton's Bill Bradley played with the Knicks before moving to the U.S. Senate.

ENTERTAINMENT

DON'T KNOW MUCH ABOUT *Movie Monsters*

GODZILLA—TO BE PRECISE *Gojira,* the original Japanese film title—made its debut in 1954. (The Americanized version, *Godzilla, King of the Monsters,* was released in 1956.) Made in the only country to suffer an atomic bombing, this monster classic was intended as a cautionary tale of the dangers of the nuclear arms race and inspired a series of increasingly campy sequels. But is Godzilla really King of the celluloid monsters? What do you know about some other famous film frights?

1. **What was added to the American version of *Godzilla*?**
2. **Recently named the Best Monster of all time, who has terrorized Fay Wray, Jessica Lange, and Naomi Watts?**
3. **Which five classic movie monsters were featured on a series of U.S. postage stamps?**
4. **What watery monster was recently named most heart pounding movie (after *Psycho*)?**
5. **What was the first vampire movie?**

ANSWERS

1. The character of an American journalist, played by Raymond Burr, later of *Perry Mason* fame.
2. *King Kong*—star of the 1933 original and a 1976 remake that was Lange's film debut—achieved the top spot in a film magazine ranking. A third version opened in 2005.
3. Dracula (Bela Lugosi), Frankenstein (Boris Karloff), the Mummy (Karloff), the Wolf Man (Lon Chaney), and the Phantom of the Opera (Chaney) were on a series of 32-cent stamps.
4. According to the American Film Institute, the shark in *Jaws*.
5. The 1922 silent classic *Nosferatu*, based loosely on Bram Stoker's novel *Dracula* and inspiration for a recent film, *Shadow of the Vampire*.

DON'T KNOW MUCH ABOUT *Canine Stars*

TELEVISION HISTORY and a unique piece of Americana started in 1954 when television's *Lassie* premiered. TV icons Timmy (Jon Provost) and June Lockhart as Ruth Martin both joined the show in later years. Yes, Lassie was always played by a male collie, and no, Timmy never fell down a well! What else do you know about other notable canine screen stars? Take a bite out of this quick quiz.

1. **Which Hollywood icon starred in the first Lassie film, *Lassie Come Home*?**
2. **Who created that American dog, Lassie?**
3. **Fans of which famous canine star have petitioned for a stamp in his honor?**
4. **What beloved screen mutt suffers from "hydrophobia"?**
5. **St. Bernards have played two of the friendliest and scariest screen dogs. Which is which?**
6. **What famous dog reappeared on screens in the summer of 2004, thirty years after his debut?**

Answers

1. Elizabeth Taylor was featured with child costar Roddy McDowall in the 1943 film, playing Priscilla. June Lockhart of TV fame played Priscilla in a later Lassie film.
2. The classic "American" dog first appeared in a short story by British author Eric Knight, and was set in England's Yorkshire country.
3. Television's classic German shepherd, Rin Tin Tin.
4. One of our favorite movie animals, Old Yeller was star of the 1957 Disney tearjerker.
5. The lovable Beethoven and the frightening Cujo (based on a Stephen King novel).
6. The perennially popular Benji returned in *Benji Off the Leash.* The dog is actually Benji #4.

DON'T KNOW MUCH ABOUT *Children's Television*

FEW PEOPLE MAKE their mark on history as quietly as Fred Rogers, who was born March 20, 1928, and died on February 27, 2003, at age 74. *Mr. Rogers' Neighborhood* first aired in Pittsburgh in 1967 and was picked up by PBS a year later. Providing an oasis of calm in the pie-throwing world of children's television, Mr. Rogers took kids and their parents into his gentle world of real people and a make-believe kingdom for thirty-four years, until the final original show was taped in December 2001. So, neighbor, tune into this quick quiz about children's television.

1. **What was the first children's television show?**
2. **After losing his job as Clarabel Clown, an actor started his own show and it became one of the longest running series in television history. What was it?**
3. **In 1955, what legendary American show debuted on afternoon television?**
4. **Which Mr. Rogers item hangs in the Smithsonian's National Museum of American History?**
5. **What was Fred Rogers's professional background?**
6. **In November 1969, what revolutionary show debuted on U.S. Public Service television stations?**

ANSWERS

1. The frenetic Howdy Doody show aired on December 27, 1947. Its phenomenal success is credited with helping to make the future of television viable.
2. A star as the clown on *Howdy Doody,* Bob Keeshan was fired and in 1955, he created *Captain Kangaroo,* which ran on CBS until 1984.
3. *The Mickey Mouse Club.*
4. His trademark red cardigan zip-up sweater.
5. He was an ordained Presbyterian minister.
6. *Sesame Street.*

DON'T KNOW MUCH ABOUT *Cartoons*

POOR GEORGE JETSON. Entering midlife and he gets no respect! Inspired by *The Flintstones*, their prehistoric cartoon counterparts, *The Jetsons* first aired in September 1962. But the futuristic family did not make it to *TV Guide*'s list of "50 Greatest Cartoon Characters" in 2002. It's a long way from Bedrock to South Park, but cartoons have been part of Americana for a long time. Think you know your toons? Take this quick quiz.

1. **What TV classic inspired *The Flintstones*?**
2. **According to *TV Guide*, who is the greatest cartoon character of all time?**
3. **What do Barney Rubble, Bugs Bunny, and Mr. Spacely of *The Jetsons* have in common?**
4. **What famed cartoon character was once censored for her short skirts?**
5. **Where does the word "cartoon" come from?**

ANSWERS

1. Fred, Wilma, Barney, and Betty, who first appeared in September 1960, were loosely based on *The Honeymooners*, the classic Jackie Gleason comedy series.

2. Bugs Bunny.

3. The voice of Mel Blanc, the most famous voice in cartoon history.

4. Betty Boop, born August 9, 1930, the creation of cartoon genius Max Fleischer's studio.

5. It is from the Italian word for "paper."

DON'T KNOW MUCH ABOUT *The War of the Worlds*

IT MAY HAVE been the scariest Halloween prank ever played. Millions of Americans turned on a popular CBS radio show on October 30, 1938, expecting a radio drama. Instead they heard "news reports" of a meteor landing, followed by a highly realistic account of a Martian invasion, including spaceships and deadly ray guns. The result was a wave of panic in a Depression-era nation already jumpy over Hitler and war in Europe. This famed broadcast was the work of radio star Orson Welles and the Mercury Theater, which produced this live radio version of the H.G. Wells science fiction classic *The War of the Worlds*. While reports of suicides and deaths in the wake of the show were exaggerated, there is no doubt that millions thought they were hearing the real thing. What do you know about this notorious moment in media history? Tune in to this quick quiz.

1. **The radio version, presented as a simulated newscast, was set in New Jersey and New York. Where did the original H. G. Wells story take place?**
2. **Orson Welles was best known as the voice of what famous radio character?**
3. **Geologists from what university went in search of the "meteor" described on the radio?**
4. **Was the audience warned that the broadcast was a drama?**

Answers

1. The Wells novel, published in 1898, is mostly set in London and England.

2. Orson Welles was the voice of the Shadow.

3. Two professors from nearby Princeton University set off in search of the "landing site."

4. Yes. There was one brief announcement at the beginning of the show, but no other announcement was made for forty minutes. There were also no commercials aired, adding to the broadcast's realism.

DON'T KNOW MUCH ABOUT *Quiz Shows*

WE NEVER SEEM to get tired of putting the "answer in the form of a question," as the old *Jeopardy* formula puts it. New game shows come and go all the time. But America has had a love affair with the quiz show since the golden days of radio and early television. In 1958, at the peak of quiz frenzy, there were twenty-two game shows on the air. One of the most popular was *The $64,000 Question*, which debuted on June 7, 1955. But then scandal swept the industry. The fix had been in. Charismatic contestants had been provided with answers to keep them on the air. Then a disgruntled contestant blew the whistle, prompting Congressional hearings. Since you love quizzes, try this one. And remember, your answer must be in the form of a question!

"ANSWERS"

1. **Top prize on the 1941 radio program *Take It or Leave It*.**
2. **Preselected to lose *The $64,000 Question*, she won the grand prize and later fame.**
3. **The big loser in the quiz show scandal, the contestant who had captivated the nation for fourteen weeks on NBC's *Twenty-One*.**
4. **1994 film about the television scandals.**
5. **Debuted in 1964, with Art Fleming as host.**

"QUESTIONS"

1. What is $64,000?
2. Who was Dr. Joyce Brothers?
3. Who was Columbia University professor Charles Van Doren?
4. What is *Quiz Show,* directed by Robert Redford?
5. What is *Jeopardy*?

DON'T KNOW MUCH ABOUT *The Wizard of Oz*

Since Dorothy of Kansas and her magical friends appeared in 1899 in Lyman Frank Baum's *The Wonderful Wizard of Oz,* they have become American icons. Born in upstate New York, Baum worked in the theater world before moving west to run a newspaper in South Dakota and write books for children. While Baum's first two books were successful, the publisher produced his "modernized fairy tale" only after Baum agreed to pay the costs. It was an instant success and spawned a musical and two silent versions before the 1939 Judy Garland classic. For years, scholars looked for spiritual, feminist, and political messages (it was once called a "Parable on Populism") buried in Baum's work. But the book remains a children's favorite. What do you know about this American classic and its author? Follow the yellow brick road to this quick quiz.

1. **What color were Dorothy's slippers?**
2. **What must Dorothy cross to return to Kansas?**
3. **The movie scarecrow gets a diploma for brains. What does he receive in the book?**
4. **Where did the word "Oz" come from?**
5. **What 1978 film did *The Wizard of Oz* inspire?**
6. **What surprising view did Baum once express about American Indians?**

ANSWERS

1. The famed "ruby slippers" worn by Judy Garland were originally silver shoes.

2. Not the rainbow of movie fame, but a great desert.

3. His head is filled with pins and needles to make him "sharp."

4. Baum said he looked at a filing cabinet which was labeled "A-N" and "O-Z." He chose the latter, otherwise it might have been *The Wizard of An*.

5. *The Wiz*, a black musical adaptation featuring Diana Ross and Michael Jackson.

6. Baum once called for the extermination of the Indians in his Aberdeen, South Dakota, newspaper.

DON'T KNOW MUCH ABOUT *The Post-Breakup Beatles*

FOR AN ENTIRE GENERATION, April 10 marks the sad anniversary of the breakup of the Beatles. When the almost unthinkable occurred in 1970, it seemed an end of an era. But as one of the four would later sing, "All Things Must Pass." The four "lads" who changed music history went on to extend their impressive achievements in new careers. Of course, Lennon was assassinated in 1980 and George Harrison succumbed to cancer in 2001. What else do you know about the Beatles after the breakup?

1. **Who had the first solo album after the split?**
2. **Which former Beatle organized one of the first major rock concerts for charity?**
3. **Did the foursome ever reunite?**
4. **For a younger generation, which former Beatle is "Mr. Conductor"?**
5. **Which Beatle is commemorated in New York's Central Park?**

Answers

1. McCartney's solo album *McCartney*, featuring "Baby I'm Amazed" and "That Would Be Something" was released in the UK one week after the breakup was announced.

2. In 1971, George Harrison organized the Concert for Bangladesh, drawing 40,000 fans to Madison Square Garden in one of the first major charity concerts.

3. No, but Paul McCartney, Ringo Starr, and George Harrison did join to record music that was added to an unfinished Lennon song, "Free As a Bird," released in 1995.

4. Ringo Starr played the part on the popular children's show, *Shining Time Station*.

5. John Lennon is memorialized with "Strawberry Fields" near the spot where he was murdered in December 1980.

DON'T KNOW MUCH ABOUT *Young Elvis*

IN AN ERA OF Madonna, Britney, and "wardrobe malfunctions," it is hard to imagine a day when Ed Sullivan would only show young Elvis Presley from the waist up and some towns banned his live concerts because of the way he swiveled his hips as he sang. Legions of fans of the legendary Elvis remember the milestone of his first commercial recording in 1954. Since his death on August 16, 1977, America's most popular singer retains his regal status. What do you know about the man they call "The King"? Put on your Blue Suede Shoes for this quick quiz.

1. **What was Elvis's first recording?**
2. **Where was Elvis born?**
3. **What was Elvis's first television appearance?**
4. **What was his first major hit song?**
5. **What was his first movie?**

ANSWERS

1. The single "That's All Right, Mama" recorded in the famed Sun Studios in Memphis.

2. Elvis Aaron Presley was born in Tupelo, Mississippi, on January 8, 1935. His twin brother, named Jesse, was stillborn and Elvis was an only child.

3. *The Milton Berle Show* in April 1956 broadcast from the deck of an aircraft carrier. Ed Sullivan did not want Elvis on, until his appearances on other shows won huge ratings. He did three Ed Sullivan appearances, but was only censored on the third time.

4. "Heartbreak Hotel," released in January 1956, which became a Billboard Number One and Elvis's first gold record.

5. *Love Me Tender*, which opened in November 1956. He made thirty-two other films.

DON'T KNOW MUCH ABOUT *Cary Grant*

BORN IN 1904 in Bristol, England, Cary Grant became a true Hollywood idol, famous for his good looks and sophisticated screen presence. Best known for his many screwball comedies, he appeared in such classics as *Topper* (1937), *Bringing Up Baby* (1938), *The Philadelphia Story* (1940), *His Girl Friday* (1940), and *Father Goose* (1964). Grant, who died in 1986, worked with many of Hollywood's greatest leading ladies, from Mae West to Katharine Hepburn, Sophia Loren, and Ingrid Bergman. What do you know about this suave leading man who also starred in the adventure classic *Gunga Din* and once played the Mock Turtle in *Alice in Wonderland*?

1. **How many films did Cary Grant make?**
2. **For which role did Grant win an Oscar?**
3. **What was Cary Grant's real name?**
4. **How many Alfred Hitchcock movies did he appear in?**
5. **How many times was he married?**
6. **In which film did Cary Grant ever utter the oft-imitated words "Judy, Judy, Judy"?**

ANSWERS

1. Grant made his film debut in 1932, and in a career spanning nearly four decades, made seventy-two feature films. He achieved his first major success in *She Done Him Wrong* (1933), in which he starred with Mae West. He retired in 1966.

2. Twice nominated for Best Actor, he never won. In 1970, he was presented with a Special Oscar.

3. His real name was Alexander Archibald Leach.

4. Best known for comedies, he starred in the thrillers *Suspicion* (1941), *Notorious* (1946), *To Catch a Thief* (1955), and *North by Northwest* (1959), the film in which he is chased through a cornfield by a plane.

5. Grant married, in order, Virginia Cherill, Barbara Hutton, Betsy Drake, Dyan Cannon, and Barbara Harris.

6. Grant never spoke those words in a film. He does mention the name once in *Only Angels Have Wings.* Cary Grant buffs suggest that an impressionist may have been "doing" Cary Grant when Judy Garland walked into the room and he spoke the famous name three times.

DON'T KNOW MUCH ABOUT
The Rolling Stones

Hey Baby Boomers. Want to feel your age? We don't know if he has his AARP card yet, but Mick Jagger turned 60 on July 26, 2003. Born in Dartford, England, in 1943, Jagger met future band mate Keith Richards at the Dartford Maypole Country Primary School. Doesn't sound like the beginnings of the baddest boys in rock 'n' roll history. But that was the unlikely beginning of the Rolling Stones, the self-proclaimed "World's Greatest Rock and Roll Band." Named for a Muddy Waters song, the band released its first single in June 1963. What do you know about these bad boys of rock and roll? Take this quick quiz.

1. **Where was Jagger studying when he reconnected with Richards in 1960?**
2. **Who is the oldest member of the original Rolling Stones?**
3. **What was the first Stones song to hit U.S. music charts?**
4. **What was their first Number One American hit?**
5. **In a notorious *Ed Sullivan Show* appearance in 1967, what song was self-censored?**
6. **In 1969, what two tragedies struck the band?**

ANSWERS

1. The famed London School of Economics.

2. Jagger and Richards, who was also born in 1943, are actually the babies of the original band. Drummer Charlie Watts is 62, but guitarist Bill Wyman was born in October 1936! (Wyman retired from the band in 1992.)

3. A version of Buddy Holly's "Not Fade Away," which made #48 on the charts in 1964, the year they made their first U.S. tour.

4. "(I Can't Get No) Satisfaction," which topped VHS's list of 100 Greatest Rock Songs in the year 2000.

5. "Let's Spend the Night Together"; Jagger later said he mumbled the offending lyrics.

6. Original member Brian Jones was found dead in his swimming pool in July and it was ruled "death by misadventure." In August, at a free concert held at California's Altamont Speedway, a young man was stabbed to death by members of the Hell's Angels motorcycle gang which was providing security for the concert. The incident is captured in the documentary film *Gimme Shelter.*

DON'T KNOW MUCH ABOUT *Hank Williams*

THE WORD "LEGEND" gets overused these days, especially in the music industry. But in the case of Hank Williams, it is well deserved. Called the greatest country singer of all time, Hank Williams Sr. died on New Year's Day in 1953 at the age of 29 after an all-too-brief career. Born Hiram King Williams on September 17, 1923, in rural Alabama, he taught himself to play when he was given a guitar at 8 years old. Unlike other country singers, Williams wrote most of the songs he sang, and songs such as "Your Cheatin' Heart" and "Hey, Good Lookin'" helped spread country music from the rural South and Southwest to the rest of the United States. What do you know about this unique American performer?

1. **When did Williams begin his recording career?**
2. **Who else had a hit with Williams's Number One country hit "Cold, Cold Heart"?**
3. **In 1961, Williams was one of the first members of the Country Music Hall of Fame. Where is it located?**
4. **How did Hank Williams die?**
5. **What was his last recorded hit?**

Answers

1. In 1947, Williams started recording in Nashville, Tennessee, the center of country music although he had started his first band, the Drifting Cowboys, in 1937.

2. The song was a pop hit for Tony Bennett.

3. Like the famed Grand Ole Opry, the Hall of Fame is in Nashville, Tennessee.

4. On January 1, 1953, he left Knoxville, Tennessee, in a chauffeur driven Cadillac for a concert in Canton, Ohio. When the chauffeur was pulled over for speeding, the police officer looked in the backseat and noticed Williams was dead. He had suffered a heart attack.

5. Following his death, his final single shot to Number One. Ironically, it was titled, "I'll Never Get Out of This World Alive."

DON'T KNOW MUCH ABOUT *Bob Hope*

IF ANYBODY COULD be called Entertainer of the Twentieth Century, the title belonged to Bob Hope, who died on July 29, 2003, at age 100. Born Leslie Townes Hope on May 29, 1903, in Eltham, England, he moved to Cleveland with his family at the age of 4. He left England, he said, "when I found out I couldn't be king." But he became King of Comedy in a career spanning vaudeville, Broadway, radio, more than fifty films, and television. Hope became one of the world's most successful, recognizable, and beloved entertainers. Cited by Guinness as the most honored entertainer in the world, Hope received more than two thousand awards and citations for his humanitarian and professional efforts. What do you know about this legend of entertainment?

1. In his first feature film, *The Big Broadcast of 1938*, what song did Hope sing for the first time?

2. The first stop was Singapore; the last was Hong Kong. What were these Hope classics?

3. What did Hope do for the first time on Easter Sunday 1950?

4. What Hope tradition was born in May 1941?

5. What was named for Hope in May 1997?

6. What unprecedented honor did Hope receive in October 1997?

Answers

1. "Thanks for the Memory," which became his signature song.

2. The famous *Road to* movies with costars Bing Crosby and Dorothy Lamour. There were six *Road to* movies, which also included destinations in Morocco, Rio, Utopia, and Zanzibar.

3. Although he had appeared on television before, Hope made his formal debut on NBC television with the special, *Star Spangled Revue*. The formula proved extremely successful and Hope's NBC career continued for the next fifty years.

4. With a group of performers, Hope went to March Field, California, to do a radio show there. Throughout World War II, Hope's radio shows were aired from military bases and installations around the globe. Hope began a Christmas custom in 1948 by entertaining troops involved in the Berlin Airlift. In 1990, he was in Saudi Arabia entertaining the men and women of Operation Desert Storm.

5. The USNS *Bob Hope*, the first of a new class of ships named after him, was christened. The U.S. Air Force later dedicated the *Spirit of Bob Hope*. In 2001, this C–17 transported the pilots and crew of the reconnaissance plane downed in China back to Hawaii.

6. Congress made him an Honorary Veteran, the first individual so honored in the history of the United States.

DON'T KNOW MUCH ABOUT *Andrew Lloyd Webber*

THERE ARE MUSICAL theater purists who may thumb their noses at *Cats*, among the most successful of the knighted Andrew Lloyd Webber's many musicals. But *Cats* opened in New York on October 7, 1982, and went on to become the longest running and highest grossing Broadway musical ever. Then, on January 9, 2006, *Cats* was surpassed in both categories by Webber's *The Phantom of the Opera* with its 7,486th performance. In December 2003, a film version of *The Phantom of the Opera* was released, the third Webber musical to be brought to the screen. With a worldwide gross of $3.3 billion, *Phantom* is considered the highest-grossing entertainment event of all time. What else do you know about the composer who was the first man to have three musicals running on Broadway at the same time? "Don't cry for" this quick quiz.

1. **Which Webber show opened first?**
2. **Which Webber "rock opera" was greeted with protests in 1971?**
3. **Name the Webber musical based on real events in South America.**
4. **Which show was written for Webber's wife at the time in a leading role?**
5. **Which show is based on a movie ranked among the greatest films ever by the American Film Institute?**
6. **Name the two Webber shows that have won seven Tony Awards apiece?**
7. **What famed poet inspired *Cats*?**

ANSWERS

1. *Joseph and the Amazing Technicolor Dreamcoat,* a collaboration with lyricist Tim Rice based on the biblical story of Joseph and his brothers, opened in London in 1968.

2. *Jesus Christ Superstar,* which some religious groups disliked because it portrayed Jesus as a man. Since then it has been widely accepted by churches and is often performed by church groups.

3. *Evita* (1978), based on the life of Eva Perón, wife of Argentine dictator Juan Perón.

4. *The Phantom of the Opera,* which opened in 1986 and is still running, was written for Webber's wife Sarah Brightman in the role of Christine. They had married in 1984 and were divorced in 1990.

5. *Sunset Boulevard,* based on the 1950 Academy Award–winning Billy Wilder film starring William Holden and Gloria Swanson.

6. *The Phantom of the Opera* and *Sunset Boulevard.*

7. The show was based on *Old Possum's Book of Practical Cats,* poems by T. S. Eliot, better known for his modern classics "The Waste Land" and "The Love Song of J. Alfred Prufrock," neither of which is likely to become a musical.

DON'T KNOW MUCH ABOUT
Katharine Hepburn

SHE IS AN AMERICAN ORIGINAL. Born in Hartford, Connecticut, on May 12, 1907, Katharine Hepburn was the daughter of a doctor and a suffragette, both of whom encouraged her strong streak of independence. After college at Bryn Mawr, she went to Broadway and then to Hollywood. But she was no typical Hollywood starlet. Once billed "box office poison," Hepburn refused interviews, wore slacks and no makeup, and wouldn't pose for pinup pictures. With twelve Academy Award nominations and four wins, she is widely considered America's greatest film actress. Hepburn died June 30, 2003, at age 96. What else do you know about this living legend of the screen?

1. **In what film did she debut?**
2. **For which films did Hepburn win Oscars?**
3. **How many films did she make with Spencer Tracy?**
4. **Which classic novel became a 1933 Hepburn box office hit?**
5. **What famous woman's personality inspired her character in *The African Queen* (1951)?**

ANSWERS

1. A *Bill of Divorcement* (1932), in which she played the daughter of John Barrymore.

2. *Morning Glory* (1933), *Guess Who's Coming to Dinner* (1967), A *Lion in Winter* (1968), and *On Golden Pond* (1981).

3. Nine, starting with *Woman of the Year* (1942). They also began a twenty-five-year off-screen relationship that lasted until his death. She claimed never to have watched *Guess Who's Coming to Dinner* because it was his last film.

4. Louisa May Alcott's *Little Women*.

5. Hepburn modeled the missionary spinster Rose on First Lady Eleanor Roosevelt in this classic pairing with Humphrey Bogart.

DON'T KNOW MUCH ABOUT *Marilyn Monroe*

More than forty years after her death, she is still larger than life. On August 5, 1962, Marilyn Monroe died in her sleep, a suicide at age 36. Her life started out as a storybook—a girl who went from orphanages and a factory job to the top—only to end as stark tragedy. Along the way, she captivated the world, had romances with some of the world's most fascinating and powerful men, and inspired more than 300 biographies, numerous documentaries, and countless rumors and myths. What do you know about this American icon?

1. **What was Marilyn Monroe's real name?**
2. **Where was she "discovered"?**
3. **What was her first film?**
4. **After a first marriage ended in divorce, who were her famous second and third husbands?**
5. **Did Monroe ever win an acting award?**

ANSWERS

1. She was born Norma Jean Mortenson (June 1, 1926), but was baptized as Norma Jean Baker. She changed her name and hair color after signing her first film studio contract in 1946.

2. She was working on a wartime assembly line when a *Yank* magazine photographer spotted her and launched her cover girl career.

3. She had a bit part in *The Shocking Miss Pilgrim* (1947), but attracted most notice for *The Asphalt Jungle* (1950). Her film career was really launched with 1950's *All About Eve.* The movies *Gentlemen Prefer Blondes* and *How to Marry a Millionaire* made her the most famous "blonde bombshell" in the world by age 27.

4. Married at age 16 to a neighbor whom she divorced, she married baseball superstar Joe DiMaggio in 1954, divorcing after nine months. In 1956, she married playwright Arthur Miller. Miller's play *After the Fall* is based on their lives.

5. Yes. She won a Golden Globe for Best Actress in a Comedy for the classic *Some Like It Hot.* Her last completed film was *The Misfits* (1961), also the last film of costar Clark Gable.

AND MORE!

DON'T KNOW MUCH ABOUT *Teeth*

IT'S NO ACCIDENT that October—with Halloween at its end—is also National Dental Hygiene Month. Guess what? We don't like going to the dentist. Besides the obvious reasons for not going, millions of Americans can't get care because there is a shortage of dentists in many places. But dental care is not just about shiny whites. Research shows gum disease as a risk factor for heart disease and other systemic diseases. Routine exams can uncover symptoms of diabetes, osteoporosis and low bone mass, HIV, and other problems. Taking care of one's teeth has been a human preoccupation since ancient times, and the Egyptians, Chinese, and Greeks all had methods for cleaning teeth and freshening breath! What do you know about dentistry? Bite down on this quick quiz.

TRUE OR FALSE?

1. **Dental caries or cavities are the most common chronic disease among young people.**
2. **George Washington's dentist invented the dental drill.**
3. **Since ancient times, teeth have been filled with stone chips and other items.**
4. **The first silver fillings were made from coins.**
5. **False teeth were invented in the 1700s.**
6. **Nitrous oxide, or "laughing gas," is a twentieth-century invention.**

ANSWERS

1. True. Nationally, cavities affect 42 percent of people 5 to 17 years, according to the CDC.

2. False. While John Greenwood is credited with the first known "dental foot engine" made from his mother's spinning wheel, the drill dates to Pierre Fauchard in 1728. A hand-cranked drill was patented in 1838 and the first motor-driven drill appeared in 1864; electricity was added in 1874.

3. True. Lead, cork, and tinfoil were also used in the nineteenth century until gold leaf became popular in the U.S.

4. True. A French dentist used fillings from silver coins mixed with mercury. This amalgam was standardized in 1895.

5. False. The Etruscans of Italy made false teeth of ivory and gold as early as 700 BC. Washington's many sets included the commonly used hippopotamus and elephant ivory and even human teeth, either pulled from the dead or sold by the poor.

6. False. The use of nitrous oxide in surgery began in the early 1800s and it became practical by 1863.

DON'T KNOW MUCH ABOUT *Wall Street*

MUTUAL FUNDS, IRAS, and 401k plans have revolutionized the way Americans invest their savings. But most of us don't know much about the stock market, even though it is a driving force in the American economy. How much do you know about the "bulls and bears"? Ring the opening bell in this quick quiz.

1. **What is the "Big Board"?**
2. **What is the Dow Jones Industrial Average (DJIA)?**
3. **Which of these companies is not part of the Dow Jones Industrial Average: Apple, Texaco, and Bethlehem Steel?**
4. **How many companies trade their stock on the New York Stock Exchange?**
5. **What is NASDAQ?**

ANSWERS

1. The nickname of the New York Stock Exchange, the oldest and largest stock exchange in the U.S.

2. The hundred-year-old index is the most famous yardstick of the American economy. Originally comprised of twelve stocks, the current DJIA is made up of the average of stock prices of thirty prominent companies traded on the New York Stock Exchange, including American Express, Boeing, Coca-Cola, General Motors, and Walt Disney. The sum of the prices of these stocks is divided by a complex formula that reflects other changes in the values of the stocks.

3. None of these companies was part of the DJIA 30 in April 2008.

4. As of December 2007, the NYSE listed more than 2,800 issuers.

5. The National Association of Securities Dealers Automated Quotation (NASDAQ) system is another major stock exchange. It has gained prominence because many leading technology companies trade their stocks through NASDAQ.

DON'T KNOW MUCH ABOUT *Historically Black Colleges*

FROM A SMALL SHANTY and church in rural Alabama, the visionary African-American educator Booker T. Washington (1856–1915) built the future Tuskegee University, beginning his efforts in 1881. While not the oldest of the nation's historically black colleges and universities, Tuskegee is among the most prominent of these schools, most of which opened in the wake of the Civil War to educate emancipated slaves. What do you know about these institutions and their proud tradition in American history? Enroll in this quick quiz.

1. **Founded after the Civil War, which college was named for a Civil War general?**
2. **Which historically black institution did Booker T. Washington attend?**
3. **What noted scientist joined Tuskegee as an instructor in 1910?**
4. **Who were the "Tuskegee Airmen"?**
5. **What is the oldest university in Nashville, whose graduates include W. E. B. Du Bois, one of the founders of the NAACP?**
6. **In 1901, which school was founded as the Colored Industrial and Agricultural School?**
7. **Originally founded in Augusta, Georgia, which school's alumni include Martin Luther King Jr. and filmmaker Spike Lee?**

ANSWERS

1. Washington, D.C.'s Howard University, the largest historically black university, is named for General Oliver O. Howard who also led the Freedmen's Bureau after the war.

2. Born a slave, Washington walked 500 miles to attend Virginia's Hampton Normal and Agricultural Institute; founded in 1868, it is now Hampton University.

3. Agricultural science pioneer George Washington Carver (1864–1943).

4. The first black men to qualify as military pilots during World War II. Tuskegee produced nearly 1,000 African-American pilots during the war, at a time when the American armed services were still segregated.

5. Fisk University, founded in 1866.

6. Louisiana's Grambling State University.

7. Atlanta's Morehouse College, founded in 1867.

DON'T KNOW MUCH ABOUT *Centenarians*

A HUNDRED YEARS may seem like a long time, but more of us are living to that ripe old age. When the *Today* show's Willard Scott first began announcing centenarian birthdays on television, he received a trickle of letters; by 2005 he was getting hundreds of 100+ birthday announcements every week. According to the Census Bureau, in 2005 there were 76,000 Americans who hold the distinction of being centenarians, now believed to be the fastest growing American age group. In 2010, the number of centenarians is projected to be 129,000. What do you know about some famous American "centurions"?

1. **What centenarian sisters wrote a bestseller and were the subject of a Broadway play in the 1990s?**
2. **What immigrant, who died at age 101, wrote dozens of the best loved songs of the twentieth century?**
3. **What cigar-chomping comedian said, "I would go out with women my age, but there are no women my age"?**
4. **What celebrated folk artist illustrated "'Twas the Night Before Christmas" at age 100?**

ANSWERS

1. The Delany sisters, Sadie and Bessie, who wrote the autobiographical bestseller *Having Our Say* after hitting 100.
2. Irving Berlin.
3. George Burns.
4. Grandma Moses, who didn't take up painting until she was in her 70s.

DON'T KNOW MUCH ABOUT
The American Red Cross

THE AMERICAN RED CROSS has roots that stretch back to the nineteenth century. But in 1905, Congress gave the group a federal charter to provide disaster relief, and March is traditionally designated as Red Cross Month. Today the American Red Cross is a familiar part of a network of societies in more than 175 nations dedicated to relieve human suffering during times of disaster. What else do you know about this familiar group that has been occasionally stung by controversy?

1. **How many disasters does the American Red Cross deal with?**
2. **How big is the nation's blood supply?**
3. **Who founded the Red Cross?**
4. **What is the most deadly American disaster?**
5. **What is the Red Crescent Society?**

ANSWERS

1. In 2007, the number was more than 72,000; on average the Red Cross responds to more than 70,000 disasters every year, including house fires, chemical spills, tornadoes, floods, and hurricanes. Mostly volunteers, Red Cross workers provide emergency food, clothing, shelter, health care, and counseling, usually within hours of the emergency.

2. According to the American Association of Blood Banks, an average of 38,000 units of red blood cells are needed daily. The primary supplier of blood and blood products in the United States, the Red Cross collected from approximately 4 million volunteers in 2007.

3. The Red Cross was founded in 1863 by Jean Henri Dunant, a Swiss philanthropist, who saw a field in Italy the day after 40,000 people had been killed or wounded in a battle. Horrified at the suffering, he formed a group of volunteers to help them. The American Red Cross was begun by Civil War nurse Clara Barton who worked to establish the American Association of the Red Cross in 1881. Barton ran the organization until she was forced out in 1904 in a controversy over funds, and the Red Cross was reorganized under its present charter in 1905 which established the basic organization of today's American Red Cross.

4. The highest natural disaster death toll in U.S. history was caused by the Galveston, Texas, hurricane of September 8, 1900, which killed an estimated 8,000 people. This disaster was the subject of the bestselling book *Isaac's Storm* by Erik Larson.

5. Societies in most Muslim countries use a red crescent on a white field and call themselves Red Crescent societies. In Israel, the society is called the Magen David Adom and has a red star of David on a white field for its flag.

DON'T KNOW MUCH ABOUT *National Anthems*

CONSIDERED BY SINGERS to be one of the world's most difficult songs, "The Star-Spangled Banner" features lyrics by Francis Scott Key during the War of 1812. A young attorney, Key was held aboard a British warship as Baltimore's Fort McHenry was bombarded on the night of September 13–14, 1814. Inspired by the flag flying at "dawn's early light," Key wrote the words for which he is immortalized. While our national hymn and "O Canada," which Americans get to hear at sports events, are familiar songs, the anthems of other nations are a mystery. What do you know about America's anthem and those of some other nations? Hum along with this quick quiz.

TRUE OR FALSE?

1. **While Francis Scott Key wrote the words, famed composer Stephen Foster wrote the music for America's anthem.**
2. **"The Star-Spangled Banner" was designated the national anthem by Abraham Lincoln during the Civil War.**
3. **The "Himno Nacional" of Mexico begins "Mexicans, when the war cry is heard, have sword and bridle ready."**
4. **Opera composer Giuseppe Verdi wrote Italy's national anthem, "O Sole Mio."**
5. **The author and composer of England's anthem, "God Save the Queen," are unknown.**
6. **Adopted in 1795, France's "La Marseillaise" is one of the oldest anthems.**

Answers

1. False. The tune was from a popular British song, "To Anacreon in Heaven."
2. False. The official designation came in 1916 by President Wilson.
3. True. It was adopted in 1864.
4. False. Goffredo Mameli wrote the words to Italy's anthem, which begins "Italian Brothers, Italy has awakened."
5. True. The earliest copy of the words appeared in a magazine in 1745.
6. True. The song was popularized during the French Revolution.

DON'T KNOW MUCH ABOUT *Uncle Tom's Cabin*

"SO YOU'RE THE little lady who made this great war." That was how President Lincoln supposedly greeted Harriet Beecher Stowe when they met during the Civil War. From a New England abolitionist family, Harriet Beecher Stowe changed history 150 years ago with the publication of *Uncle Tom's Cabin*. The book did what politicians could not—put a human face on slavery. The bitterness aroused by Stowe's tale of slaves and their owners helped bring about the Civil War. Little read and largely misunderstood today, *Uncle Tom's Cabin* is one of America's most important books. What do you know about the "little lady" who wrote it? Try this quick quiz.

1. **What infamous law inspired Stowe to write the book?**
2. **What was the reaction to the book?**
3. **Who is Uncle Tom?**
4. **Who is Simon Legree?**
5. **How did the phrase "Uncle Tom" become an insult?**

ANSWERS

1. Passage of the Fugitive Slave Act of 1850, which compelled Americans to turn in runaway slaves. The harsh law was even condemned by those who were not abolitionists.

2. It became an overnight bestseller around the world. But many southerners despised Stowe and some even mailed her the ears of slaves.

3. The title character is a dignified black slave and the book describes his experiences with three slaveholders after he is sold away from his family.

4. The book's chief villain, Legree, is a northern slave-overseer who beats Tom when he refuses to tell where escaped slaves are hiding.

5. Crude "Tom Shows," which played in small towns, distorted the original story and characters, so the term "Uncle Tom" came to stand for a black man who adopts a humble manner to gain favor with whites. Uncle Tom was actually a heroic man who dies rather than submit.

DON'T KNOW MUCH ABOUT *Boy Scouting*

IT'S NOT JUST about helping old ladies across the street and rubbing two sticks together anymore. Through their "Good Turn for America" Program, Boy Scouts are turning the old "good turn each day" into thousands of volunteer hours at homeless shelters and other community service organizations. "On your honor," what do you know about the Boy Scouts, celebrating 100 years in February 2010?

1. **What is the "entry level" for Boy Scouts?**
2. **What is a Webelo?**
3. **What is a Jamboree?**
4. **Which are the most popular merit badges?**
5. **What do Donald Rumsfeld, Bill Bradley, and Ross Perot have in common?**
6. **Who chartered the Boy Scouts?**

ANSWERS

1. Cub Scouts is for boys in first through fifth grades.
2. A program for 10-year-old boys, it means "WE'll BE LOyal Scouts."
3. Since 1937, Boy Scouts of America has held a large encampment for Scouts every four years. The first was at the Washington Monument. The next, in 2010, will celebrate the centennial of the Boy Scouts of America at Fort A. P. Hill in Virginia.
4. First Aid is most popular, followed by Swimming.
5. They are Eagle Scouts, the highest advancement rank in scouting.
6. Chicago publisher William D. Boyce. While visiting England, Boyce was lost in the fog and was aided by a young Scout who then refused a tip from Boyce. Impressed, Boyce sought out the founder of the English Boy Scouts, Robert Baden-Powell. When Boyce returned to America, he established the American group.

WHAT ELSE DO YOU KNOW about the Boy Scouts? Try the bonus quiz.

TRUE OR FALSE?

1. The English Boy Scouts were founded in 1908 by Sir Robert Baden-Powell, who was inspired by American Indians he saw on a hunting trip.

2. *The Handbook for Boys,* the official scout's guide, was first published in 1927 and has outsold the Holy Bible.

3. Every Boy Scout promises to keep himself "physically strong, mentally awake and morally straight."

4. The Scout slogan is "Do a good turn daily."

ANSWERS

1. False. Sir Robert Baden-Powell was inspired by African scouts he saw during the Boer War.
2. False. The Bible has always been the most popular book in America, but according to the Boy Scouts, nearly 38 million copies of the Boy Scout's handbook are in print, putting it among the all-time bestselling books in America.
3. True, according to the Scout Oath.
4. True. "Be prepared" is the Scout motto.

DON'T KNOW MUCH ABOUT *Paperback Books*

AS REVOLUTIONS GO, this one was tame. But in 1935 in England, Penguin Books was born to produce inexpensive, paperbound books. In America, Pocket Books—featuring a motherly kangaroo named Gertrude on the spine—followed in 1939 and a revolution in America's reading habits began. Before the paperback appeared, there were few bookstores in America and most Americans didn't own a book, other than a Bible. Overnight, the paperback changed that forever. The mass market paperback, sold in drugstores and newsstands where books had never been carried, made America a nation of readers as never before. The original 25-cent cover price is long gone, but what do you know about this modern marketing phenomenon?

TRUE OR FALSE?

1. **When introduced to America by Pocket Books, all paperbacks cost a quarter.**
2. **Pocket Book #1 was James Hilton's novel of Shangri-La, *Lost Horizons*.**
3. **During World War II, soldiers received millions of free paperbacks.**
4. **The best-selling paperback book of all time is *Gone With the Wind*.**
5. **In 1953, paperbacks were investigated by Congress as being part of the Communist conspiracy.**

ANSWERS

1. True. Twenty-five cents, or "two-bits," was the going price. Eventually longer novels and nonfiction books were priced at 35 cents.

2. True and False. A market test had been done with Pearl Buck's novel of China, *The Good Earth,* but Hilton's novel was officially Pocket Book #1 in an initial list of ten books.

3. True. Known as Armed Services Editions, these free books were distributed to GIs through the war years. Their motto was, "Books Are Weapons in the War of Ideas."

4. False. While the combined sales of Bibles and annual books (such as the *Guinness Book of World Records* and *World Almanac*) are higher, Dr. Benjamin Spock's *Common Sense Book of Baby and Child Care,* first published in 1946 in time for the wave of baby boomers, is considered the single bestselling paperback book. In its several revised editions, the paperback has estimated sales of more than 60 million copies. The bestselling book of all time by a single author in various editions is thought to be the famous *Little Red Book* or *Quotations from Chairman Mao Zedong,* first published in 1964 and which has sold more than an estimated 1 billion copies. During Chairman Mao's life, Chinese citizens were required to own, read, and carry it at all times. Failure to produce it, especially during the period known as the Cultural Revolution, could result in beatings or a prison term.

5. False. They were investigated for their often lurid cover art and provocative subject matter and were accused of being pornographic.

DON'T KNOW MUCH ABOUT *January*

HAVE YOU BROKEN any resolutions yet? January is the month of fresh starts, since the Romans made it their first month more than 2,000 years ago. And in case you missed it, January is, among other things, National Blood Donor, Hobby, Hot Tea, Oatmeal, and Soup month. What else do you know about the fresh page on the calendar?

1. **The word "January" means "dark and cold."**
2. **The Anglo-Saxons called the first month "Wolf-month."**
3. **January's full moon is called the "First Moon."**
4. **January's birthstone is the diamond because it looks like ice.**
5. **The coldest day in America came in January.**

Answers

1. False. The month is named for Janus, the Roman god of beginnings and keeper of the gates of heaven. He had two faces and could look forward and backward. The Romans made January the first month, and in 46 BC, Julius Caesar added a day to January, making it thirty-one days long.

2. True. It was the time when wolves came into villages in search of food.

3. False. In Native American tradition it is called the "Wolf Moon" for the same reason the Anglo-Saxons called it Wolf-month. In other traditions it is the "After Yule Moon" or "Old Moon."

4. False. It is the garnet, which comes from a Latin word meaning pomegranate for the stone's resemblance to the fruit. Traditionally, garnets have been thought of as protective and healing stones.

5. True. The record cold of -80°F was recorded in Alaska in January, 1971. The world's record cold was recorded in Antarctica in July 1983; in the Southern Hemisphere, the seasons are reversed and it is winter in July.

DON'T KNOW MUCH ABOUT *Banned Books*

IN THE 1930S, books deemed dangerous by Germany's Nazi party—including works by Freud, Steinbeck, and Hemingway—were thrown into bonfires. But you don't need to burn a book to censor it. That's why the American Library Association and several other groups sponsor "Banned Books Week." The week celebrates the Freedom to Read. Each year, the Library Association releases lists of books that are frequently "challenged" by groups, usually parents who want books they deem "offensive" removed from schools and public libraries. Don't know much about banned books? As the motto of Banned Books Week puts it, "It's your freedom we're talking about."

1. **Which American President once said, "Don't join the book burners"?**

2. **Which frequently "challenged" author said, "I've met thousands of children and not even one time has a child said, 'I'm so glad I've read these books because now I want to be a witch'"?**

Answers

1. Dwight D. Eisenhower in a commencement address at Dartmouth in 1953. He added, "Don't be afraid to go in your library and read every book."
2. Harry Potter series author J. K. Rowling, whose books have been among the most challenged titles in schools and libraries.

DON'T KNOW MUCH ABOUT *The U.S. Air Force Academy*

President Eisenhower signed the U.S. Air Force into existence on April 1, 1954. What do you know about the service academy that prepares young men and women for careers as officers in the U.S. Air Force? Take off with this quick quiz.

True or False?

1. **The Air Force Academy is located an on air base near Denver.**
2. **Students entering the academy make a twelve-year commitment.**
3. **Academy students are officially known as "wingmen."**
4. **Tuition at the academy is paid by the government.**
5. **The academy has admitted women since its founding.**

Answers

1. False. The first class of 306 cadets began training in 1955 at a temporary site at Lowry Air Force Base, near Denver. But in 1958, the academy moved to its permanent site in the foothills of the Rocky Mountains, near Colorado Springs, Colorado.

2. True. Students who enter the academy agree to serve four years as a cadet and eight years as a commissioned officer, with at least five years of active Air Force duty.

3. False. Students are called Air Force Academy cadets and take four years of academic work leading to a Bachelor of Science degree along with military training to earn a commission in the U.S. Air Force.

4. True. The cost of an Academy education—$35,564 per year in 2007—is paid by the United States government, which also provides food, housing, and medical care for the cadets. Each cadet also receives a monthly salary to pay for uniforms, textbooks, and personal expenses. Cadets who fail to complete any period of active duty may be required reimburse the U.S. government for an appropriate proportion of their academy education costs.

5. False. The academy admitted women students for the first time in 1976.

DON'T KNOW MUCH ABOUT *Punctuation*

WHEN NATIONAL PUNCTUATION DAY rolls around this year (it's often in September at the start of school), make sure you know where your apostrophe is! Punctuation is simply the use of certain marks and spaces in writing and printing to make the writer's meaning clear. Although the Greeks and Romans used a few marks, most punctuation has developed since the printing era began with Gutenberg, around 1440. In the modern era of e-mail and text messaging, proper punctuation is becoming a lost art. Don't know a comma from a colon? Punctuate this quick quiz.

1. **What is the most common punctuation mark?**
2. **What was the most costly punctuation error in history?**
3. **What is an "interrobang"?**
4. **What is a "greengrocer's apostrophe"?**
5. **When the British come to an end of a sentence, what punctuation do they use?**
6. **If you ask or exclaim in Spanish, what do you add to a sentence?**
7. **What handbook of punctuation became a surprise international bestseller?**

ANSWERS

1. The comma. It has more uses than any other mark of punctuation.

2. Probably the comma left out of the computer program for NASA's *Mariner I* Venus probe, which resulted in the craft's destruction.

3. In the early 1960s, publishers tried to introduce a mark that combined a question mark and an exclamation point. Though it was added to some typewriter keyboards, it never caught on.

4. Wrongly placed apostrophes that occur frequently in handwritten signs for produce such as "potatoe's." A simple plural does not require an apostrophe, which is properly used to denote the possessive.

5. It is called a "full stop" rather than the American period.

6. An inverted question mark or exclamation point at the beginning of the sentence.

7. *Eats, Shoots & Leaves* by Lynne Truss.

DON'T KNOW MUCH ABOUT *The* NAACP

ON FEBRUARY 12, 1909—the hundredth anniversary of Abraham Lincoln's birth—a group of black and white activists created what was first called the National Negro Committee. It later became the National Association for the Advancement of Colored People, or NAACP, and the organization has been at the forefront of the civil rights struggle ever since. In the decades since its founding, the NAACP has played a key role in many of the milestones of the American quest for racial equality. What do you know about this important institution?

1. **Who was W. E. B. Du Bois?**
2. **What action taken by President Wilson sparked one of the first NAACP actions?**
3. **Which NAACP field director was assassinated in 1963?**
4. **Who served as secretary of the Montgomery, Alabama, branch of the NAACP after joining in 1943?**
5. **What case did NAACP attorney Thurgood Marshall successfully argue before the Supreme Court in 1954?**

Answers

1. A founding member of the NAACP, Du Bois (pronounced doo BOYS) was the first black man to receive a Ph.D. from Harvard. A famed historian and sociologist, his ideas appeared in a collection of essays called *The Souls of Black Folk* (1903). From 1910 to 1934, he was the editor of the NAACP magazine, *The Crisis.*

2. In 1913, Wilson officially introduced segregation into the federal government and the NAACP launched a public protest.

3. Medgar Wiley Evers, who fought against segregation and racial discrimination in Mississippi during the 1950s and early 1960s. Evers was shot and killed outside his home in Jackson, Mississippi, on June 12, 1963. Police arrested white supremacist Byron De La Beckwith for the crime, and his fingerprints were on the rifle that killed Evers. But the charges against Beckwith were dropped in 1969. In 1989, the case was reopened after new evidence was found and a jury of eight African Americans and four whites convicted Beckwith of Evers's murder in 1994.

4. Rosa Parks was an active NAACP member and chapter secretary when she was arrested for refusing to give up her bus seat in 1953, sparking a boycott and setting the nonviolent civil rights movement in motion.

5. The landmark school desegregation case, *Brown v. Board of Education.* Marshall later became the first black member of the Supreme Court.

DON'T KNOW MUCH ABOUT *Family Vacations*

AMERICANS LOVE VACATIONS, but aren't very good at taking them. Two weeks used to be the rule—half what the typical European enjoys. But hardworking Americans are cutting back vacation time by 10 percent according to travel experts. Once the exclusive province of the rich who went to spas and resorts such as Saratoga and Newport, Rhode Island, the vacation became more democratic after the Civil War. Then trains and automobiles changed everything. But where do we like to go? And are we there yet?

1. **Where was America's first Ferris wheel and amusement midway built?**
2. **Where was America's first roller coaster?**
3. **What was America's first designated national park?**
4. **Where do you go for America's largest national park?**
5. **If you went to Rehoboth, Delaware, or Ocean Grove, New Jersey, in the nineteenth century, what would you do?**

ANSWERS

1. Chicago, for the 1893 Columbian Exposition, or World's Fair. Its success inspired the first modern amusement park on Chicago's south side.

2. "Leap the Dips" in Altoona, Pennsylvania's Lakemont Park opened in 1902.

3. Yellowstone, shared by Wyoming, Montana, and Idaho, is the world's first national park, with the land set aside in 1872.

4. Alaska's Wrangell–St. Elias, with 8.3 million acres, is the largest area in the National Park system. In the lower forty-eight, try Death Valley (in California and Nevada), but maybe not in the summer. With the lowest point in America, it is one of the hottest places in the nation.

5. Pray and sing hymns. Both began as Methodist revival camps where alcohol and Sunday swimming were forbidden.

DON'T KNOW MUCH ABOUT *August*

THE RODNEY DANGERFIELD of the calendar, August gets no respect. It has no real holidays to speak of. Poets rarely praise it. It is a long month (more about that later) that leads inevitably back to school. What do you know about this often-overlooked month?

1. **How did August get its name?**
2. **Why does it have thirty-one days?**
3. **Why are the "dog days" of summer in August?**
4. **What were *The Guns of August*?**

ANSWERS

1. It is named for the Roman Emperor Augustus Caesar. Born Gaius Octavianus, he was the nephew of Julius Caesar and became Rome's first emperor in 29 BC. He was renamed *Augustus*, meaning "imperial majesty," and what was then the sixth month of the year (*Sextillus*, in Latin) was renamed in his honor because he considered it his lucky month.

2. Not to be outdone by his uncle, for whom July is named, Augustus stole a day from February so that his month was just as long as the one named for Uncle Julius. These are the only two months named after actual people and one more proof that it is good to be the emperor.

3. Mad dogs and Englishmen may go out in the noonday sun, but that has nothing to do with this phrase. The expression comes from ancient Rome where the rising of the Dog Star, Sirius, was connected to the sultry summer heat of July and August.

4. The title of Barbara Tuchman's classic history of the days in August 1914 leading up to World War I.

DON'T KNOW MUCH ABOUT
The European Union

WHEN SWEDES VOTED in 2003 to reject the euro as their official currency, it was a sharp blow to the European Union. Born on November 1, 1993, the European Union grew out of a postwar dream to end the cycle of bloody wars that had torn Europe for centuries, including the two world wars of the twentieth century. What began as the six-member European Economic Community has grown to a continental group stretching from the Arctic Circle to the Mediterranean, with a Parliament, central bank, and court system. While each member nation remains independent, the European Union acts as a single market in which goods, services, people, and capital can move freely about Europe, with customs and passport checks virtually abolished. What else do you know about this continental economic giant?

1. **How many nations make up the European Union?**
2. **How many members use the Eurodollar or euro as currency?**
3. **What is the population of the European Union countries?**
4. **Does the European Union have an anthem? A flag?**

ANSWERS

1. As of November 2007, there were twenty-seven E.U. members. The original twelve who joined in 1993 were Belgium, Denmark, France, Germany, Greece, Ireland, Italy, Luxembourg, the Netherlands, Portugal, Spain, and the United Kingdom. In 1995, Austria, Finland, and Sweden joined. Cyprus, Czech Republic, Estonia, Hungary, Latvia, Lithuania, Malta, Poland, Slovakia, and Slovenia joined in 2004. Bulgaria and Romania joined in 2007. Croatia, Turkey, and Macedonia (the former Yugoslav republic) are candidates in negotiations to join.

2. By 2008, fifteen E.U. countries used the single currency known as the euro. Officially known as the Eurozone, but also called "euroland," those countries are Austria, Belgium, Cyprus, Finland, France, Germany, Greece, Ireland, Italy, Luxembourg, Malta, the Netherlands, Portugal, Slovenia, and Spain. In 2008, twelve E.U. countries—including Denmark, Sweden, and the United Kingdom—do not use the euro, though some of these will eventually join the Eurozone.

3. According to the E.U., the combined population of the twenty-five-member countries on January 1, 2004, was more than 490 million, which would make it the third largest population after China and India.

4. Yes. The anthem is based on Beethoven's "Ode to Joy" from the Ninth Symphony. The flag is a blue field with a circle of twelve stars; the number twelve traditionally represents perfection and harmony. However, all member nations retain their national flags and anthems.

DON'T KNOW MUCH ABOUT *The Marines*

THEY HAVE BEEN CALLED a "few good men" (and women). On November 10, 1775, the Continental Congress established a marine corps as a special armed amphibious fighting group and they made their first landing in the Bahamas in 1776. Although disbanded after the Revolution, the Marine Corps was re-created as a military service by Congress in 1798. Since then, Marines have often been the first to fight in almost every major American conflict. What else do you know about these "leathernecks"? Enlist in a quick quiz or drop down and give me twenty-five pushups.

1. **What are the "Halls of Montezuma" in the Marine Hymn?**
2. **When did Marines go to the "Shores of Tripoli"?**
3. **What does the expression "Leatherneck" mean?**
4. **Army officer Robert E. Lee led marines into what significant battle?**
5. **What does the expression "Semper Fi" mean?**

Answers

1. During the Mexican War, from 1846 to 1848, Marines entered Mexico City and raised the American flag over the National Palace, which later became known as the "halls of Montezuma."

2. In 1805, Marines led the storming of a Barbary States' stronghold at Derna, Tripoli, helping end pirate raids on ships in the Mediterranean Sea.

3. The nineteenth-century Marines who battled pirates in Tripoli wore a high leather collar to ward off sword blows to the neck; hence the nickname "leatherneck."

4. When John Brown and his followers captured the army arsenal at Harpers Ferry, Virginia, in 1859 before the Civil War, Marines from Washington were the only troops available. Led by Lee, they captured Brown and occupied the arsenal.

5. It is shorthand for the Marine motto "Semper Fidelis," Latin for "always faithful."

DON'T KNOW MUCH ABOUT *Twins*

DON'T RUB YOUR EYES if you happen to be in Twinsburg, Ohio. You are not seeing double. It might be Twins Days, an annual event that attracts twins by the thousands to this town just south of Cleveland on the first weekend in August. (No, you don't have to dress alike, say organizers.) Through history, twins have been the source of considerable fascination and myth. What do you know about this genetic double take?

TRUE OR FALSE?

1. Twins are literally "one in a million," occurring once in every million births.

2. Identical twins are always the same sex.

3. Fraternal twins are equally common among all ethnic groups.

4. Laura Lee Hope, creator of the Bobbsey Twins, and Franklin Dixon, who created the Hardy Boys, are the same person.

5. The constellation Gemini is supposed to be the legendary twins Romulus and Remus, founders of Rome.

6. Ishmael and Isaac, the sons of Abraham, were a set of biblical twins.

Answers

1. False. Twins occur about once in every 89 births. By contrast, triplets occur about once in every 7,900 births, and quadruplets occur about once in 705,000 births.

2. True. Identical twins, originating in a single egg, are always of the same sex and have an identical genetic makeup. Fraternal twins, born from two separate fertilized eggs, may be of the same sex or consist of a brother and a sister, with each individual having a different genetic makeup.

3. False. The rate at which identical twins occur—about 4 times in every 1,000 births—is fairly constant. But fraternal twins occur more frequently among black people, particularly black Africans, than among people of European descent, including white Americans. Fraternal twins are least common among Asians.

4. True. Edward Stratemeyer (1862–1930) created some of the most popular characters in children's literature. He founded the Stratemeyer Syndicate which employed a staff to write the books. His other pen names include Victor Appleton for Tom Swift books and Carolyn Keene for the Nancy Drew series.

5. False. Castor and Pollux were twin heroes in Greek mythology placed together in the sky as the constellation Gemini, or The Twins.

6. False. Esau and Jacob, the sons of Isaac, were twins. Ishmael was Isaac's older half-brother.

DON'T KNOW MUCH ABOUT *State Fairs*

PIG RACES, CABBAGE WEIGH-INS, corn dogs, and tractor pulls. It must be State Fair time! An end-of-summer rite, state fair season kicks in around late August and lasts through the fall, when the self-proclaimed largest state fair in—where else?—Texas runs through October. Originally meant to highlight the various states' products, many states still hold fairs that emphasize agriculture and livestock. But auto and horse racing, concerts and the Midway with Ferris wheels, roller coasters, and cotton candy are the big draws now. What do you know about these pieces of Americana?

1. **Which state lays claim to holding the first state fair?**
2. **Which state once had the theme "2001: A Moose Odyssey"?**
3. **What do the four H's in the 4-H Club, a staple at most state fairs, stand for?**
4. **More than 3 million people will pass by the famed symbol of the Texas state fair. Who is he?**
5. **Set at Iowa's state fair, a 1932 novel inspired what famous musical duo?**

Answers

1. New York, which had its first state fair in 1841. It now draws more than a million people to the Syracuse fairgrounds.

2. Alaska.

3. Head, Heart, Hands, and Health. The 4-H Club began informally more than a hundred years ago to promote agricultural education. It is now a broader service organization dedicated to young people.

4. Big Tex, the icon of the Texas fair, who stands fifty-two feet tall and began greeting fairgoers in 1953.

5. Richard Rodgers and Oscar Hammerstein who wrote *State Fair,* their only original film score, for a 1945 movie.

DON'T KNOW MUCH ABOUT *The Mormons*

WHEN BRIGHAM YOUNG (1801–1877) led the first Mormon pioneers to Utah in 1844, he may not have envisioned an Olympic Village rising out of the desert one day. But the charismatic and controversial Young became one of the most significant colonizers of the West. Persecuted for its beliefs and practices, the Mormon church—the name commonly given to the Church of Jesus Christ of Latter-day Saints—might not have survived if not for Young, who eventually led 100,000 followers west to Utah. What else do you know about one of America's "homegrown" religious groups and Brigham Young?

TRUE OR FALSE?

1. Joseph Smith, who established the Mormon church in 1832, was murdered by an anti-Mormon mob in 1844.

2. The Book of Mormon, which Joseph Smith said contains divine revelations, was discovered near the Dead Sea.

3. Hundreds of people were killed in the Mormon War, a conflict between the U.S. government and the Mormons.

4. One of the most controversial Mormon doctrines, polygamy, or having more than one wife, was abolished in 1890.

Answers

1. True. A mob in Illinois, which resented Mormon ideas and practices, killed Smith and his brother. Brigham Young then became leader and took the group west.

2. False. Joseph Smith said he was given the Book of Mormon by an angel named Moroni in upstate New York. According to Smith, the book is a history of America's ancient inhabitants who were descended from the Israelites. It was written on golden plates in an unknown language, which Smith translated and published in 1830.

3. False. Reports of a rebellion by Mormons led to a conflict in 1857–1858, also called the Utah War. President James Buchanan sent troops to Utah, but no one was killed.

4. True and false. In 1890, the practice was publicly abandoned when the Supreme Court ruled it illegal, and Utah was allowed to join the Union. But some Mormon leaders continued the practice secretly for another fourteen years. Brigham Young himself had two dozen wives and fifty-eight children. At least one splinter group of Mormons continues the practice, which is illegal and banned by the main body of the Mormon church.

DON'T KNOW MUCH ABOUT *The Nobel Prizes*

FOR MORE THAN 100 YEARS, the most honored prizes in the world have been awarded on December 10, the date of the death of Alfred Bernhard Nobel (1833–1896), the Swedish chemist who invented dynamite. Using profits from the manufacture of explosives, he established a fund to award annual prizes to people whose work most benefited humanity. Nobel experimented in his father's factory seeking to make nitroglycerin safe to use and named his powdery mixture "dynamite," a Swedish word derived from the Greek word for "power." Greatly interested in literature, Nobel wanted the profits from explosives to be used to reward human ingenuity. To the initial list of prizes in Chemistry, Literature, Medicine, Peace, and Physics first awarded in 1901, a prize in economics was added in 1969. Can you take the prize in this quick quiz?

1. **How many American Presidents have won the Nobel Prize for Peace?**
2. **Who was the last American to win the Nobel Prize for Peace?**
3. **Who was the last American to win the Nobel Prize for Literature?**
4. **Who was the first female recipient of a Nobel Prize?**
5. **What was Einstein's Nobel Prize for?**

ANSWERS

1. Three. Theodore Roosevelt (1906), Woodrow Wilson (1919), and Jimmy Carter (2002).

2. Vice President Al Gore and the Intergovernmental Panel on Climate Change won the 2007 prize.

3. Novelist Toni Morrison, whose works include *Beloved, Song of Solomon,* and *The Bluest Eye.*

4. Marie Curie, who shared the 1903 Prize in Physics with her husband for their work on radiation. She also won the 1911 Prize in Chemistry for her work with radium. In addition, her daughter Irene Joliot-Curie won the Chemistry Prize in 1935 with her husband Frederic Joliot-Curie.

5. Not for his famous $E=mc^2$ equation as is often assumed, but for his service to theoretical physics and the discovery of the photoelectric effect.

DON'T KNOW MUCH ABOUT *The Girl Scouts*

On March 12, 1912, a group of 18 girls registered in a Savannah, Georgia, Girl Scout troop begun by founder Juliette Gordon Low. Initially based upon the British Girl Guides, the American version quickly won a devoted following. That's a lot of cookies—close to 200 million boxes worth in recent years. But this organization is about more than cookies. During the two world wars, the Girl Scouts set high standards for supporting the troops and selling war bonds. Since then, more than 50 million women have participated in Girl Scouting. In 2007, there were more than 3.7 million Girl Scouts in the United States—2.7 million girl members and more than 920,000 adult members. The Girl Scouts are also part of a worldwide organization of Girl Guides and Girl Scouts with more than 10 million members.

1. **In addition to tips on first aid, camping, and housekeeping, what did the first Girl Scout guide have instructions for?**
2. **Besides "Needlewoman" and "Cook," in which traditionally "male" activities could some early proficiency badges be earned?**
3. **What is unique about the Aerospace Proficiency badge?**
4. **What's the most popular Girl Scout cookie?**
5. **Will the cookies make me look fat?**

ANSWERS

1. Stopping runaway horses and tying up a burglar with "eight inches of cord."

2. Girl Scouts could earn the equivalent of the Boy Scout's merit badges in such activities as signaling and telegraphy.

3. It was flown aboard the space shuttle *Discovery* in April 1985. Shuttle mission specialist M. Rhea Edison was a former Girl Scout.

4. The perennial favorite is Thin Mints with 25 percent of the sales. In recent years, they were followed by Samoas (aka Caramel DeLites), Peanut Butter Patties (aka Tagalongs), Peanut Butter Sandwiches (Do-si-dos), and Shortbread/Trefoils, shaped like the Girl Scout emblem.

5. Maybe they will if you eat them by the box. But in response to obesity concerns, the Girl Scouts decided to eliminate trans fats—widely understood to be unhealthy—from their cookies by 2007.

DON'T KNOW MUCH ABOUT *The Bible*

WHETHER YOU BELIEVE in the Bible or not, it is one of the world's most influential books, with enormous impact on our laws and language. But while most Americans own a Bible, very few read it for themselves. In a time of controversy over creationism in the classroom, taking a fresh look at the Good Book may provide some interesting surprises. What do you know about the "genesis" of the Bible?

1. **Where does the word "Bible" come from?**
2. **In what language was the Bible written?**
3. **What did King James have to do with the Bible?**
4. **What are the Dead Sea Scrolls?**
5. **Where is the "Lost Ark" stored?**
6. **Who said, "To eat, to drink, and to be merry"?**

ANSWERS

1. "Bible" comes from the ancient Greek word *byblos*, for "books." That word was derived from the Phoenician city of Byblos—in modern Lebanon—where much of the papyrus used for writing was once produced.

2. The Old Testament, or Hebrew Bible, was largely composed in ancient Hebrew. There are also small sections composed in Aramaic, the common language spoken at the time of Jesus. The Christian New Testament was originally written in Greek.

3. King James was on the throne when the first authorized English translation of the Bible was issued in 1611, hence the King James Version. It remains the most familiar and popular English translation. The oldest English Bible appeared around 1384, when John Wycliffe's translation from the Latin Bible was published.

4. These ancient scrolls, dating from before the time of Jesus, include the oldest known versions of most of the books of the Old Testament, or Hebrew Bible.

5. The Ark of the Covenant, the elaborate chest that held the Ten Commandments, has never been found, in spite of the Indiana Jones movie.

6. The author of Proverbs 8:15. The prophet Isaiah later added, "for tomorrow we shall die" (Isaiah 22:13).

ACKNOWLEDGMENTS

A great many friends and colleagues helped me pull this project together, and I am very grateful to all of them. Among the many people who have been so supportive at HarperCollins, I would like to thank Carrie Kania, Jen Hart, Hope Innelli, Cal Morgan, Andrea Rosen, Michael Signorelli, Alberto Rojas, Diane Burrowes, Virginia Stanley, Nicole Reardon, and Laura Reynolds.

All of the people at the David Black Literary Agency are not only industrious, smart, and good-looking, they are also my very good friends. My life and work have been made better by David Black, Leigh Ann Eliseo, Dave Larabell, Gary Morris, Susan Raihofer, Joy Tutela, and Antonella Iannarino.

This material started out years ago with a single quiz in *USA Weekend*, where it was my pleasure to work with Jack Curry and Tom Lent, who have since moved on. My gratitude to them and the rest of the staff at the magazine.

My family is always my greatest inspiration and so my thanks and love go out to Colin and Jenny. And to my wife, Joann Davis, who makes it all possible and worthwhile.

Dorset, Vermont
March 2008

INDEX OF SUBJECTS

Famous People

Exceptional Places

Historic Happenings and Civics

Holidays and Traditions

Everyday Objects and Remarkable Inventions

Space and the Natural World

Sports

Entertainment

And More!

NEW IN HARDCOVER FROM KENNETH C. DAVIS

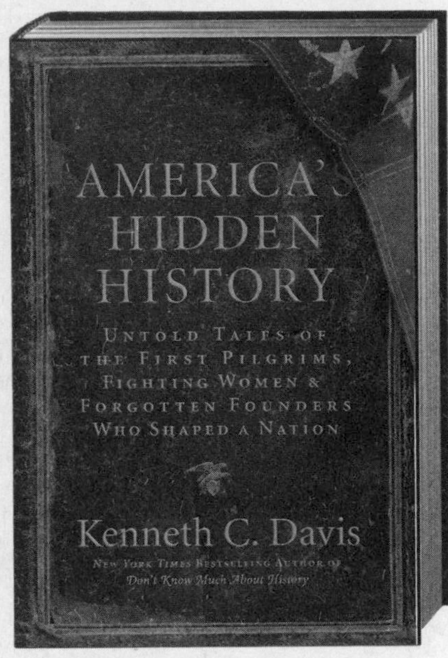

978-0-06-111818-0 (hardcover)

"History in Davis' hands is loud, coarse, painful, funny, irreverent—and memorable."
—*San Francisco Chronicle*

In this dazzling work of American history, Ken Davis—one of the country's most popular historians—presents a collection of extraordinary stories, each detailing an overlooked episode that shaped the nation's destiny and character. Spanning the period from the Spanish arrival in America to George Washington's inauguration in 1789, *America's Hidden History* details these episodes among others:

- The story of the real first Pilgrims—wine-making French Huguenots
- The coming-of-age story of Queen Isabella, who suggested that Columbus travel with the mess hall of pigs that may have spread disease to many Native Americans
- The long, bloody relationship between the Pilgrims and Indians that runs counter to the myth of the idyllic Thanksgiving feast
- The little-known story of George Washington as a headstrong young soldier who committed a war crime, signed a confession, and started a war...

Available Wherever Books Are Sold

Smithsonian Books Collins